MASSIMO DI PIERRO

ANNOTATED ALGORITHMS IN PYTHON

WITH APPLICATIONS IN PHYSICS, BIOLOGY, AND FINANCE (2ND ED)

EXPERTS4SOLUTIONS

For more information about appropriate use of this material, contact:

Massimo Di Pierro
School of Computing
DePaul University
243 S Wabash Ave
Chicago, IL 60604 (USA)
Email: massimo.dipierro@gmail.com

Library of Congress Cataloging-in-Publication Data:

ISBN: 978-0-9911604-0-2
Build Date: April 21, 2016

to my parents

Contents

6

1

Introduction

This book is assembled from lectures given by the author over a period of 10 years at the School of Computing of DePaul University. The lectures cover multiple classes, including Analysis and Design of Algorithms, Scientific Computing, Monte Carlo Simulations, and Parallel Algorithms. These lectures teach the core knowledge required by any scientist interested in numerical algorithms and by students interested in computational finance.

The notes are not comprehensive, yet they try to identify and describe the most important concepts taught in those courses using a few common tools and unified notation.

In particular, these notes do not include proofs; instead, they provide definitions and annotated code. The code is built in a modular way and is reused as much as possible throughout the book so that no step of the computations is left to the imagination. Each function defined in the code is accompanied by one or more examples of practical applications.

We take an interdisciplinary approach by providing examples in finance, physics, biology, and computer science. This is to emphasize that, although we often compartmentalize knowledge, there are very few ideas and methodologies that constitute the foundations of them all. Ultimately, this book is about problem solving using computers. The algorithms you will learn can be applied to different disciplines. Throughout history, it is not uncommon that an algorithm invented by a physicist would find application in, for example, biology or finance.

Almost all of the algorithms written in this book can be found in the `nlib` library:

```
https://github.com/mdipierro/nlib
```

1.1   Main Ideas

Even if we cover many different algorithms and examples, there are a few central ideas in this book that we try to emphasize over and over.

The first idea is that we can simplify the solution of a problem by using an approximation and then systematically improve our approximation by iterating and computing corrections.

The divide-and-conquer methodology can be seen as an example of this approach. We do this with the insertion sort when we sort the first two numbers, then we sort the first three, then we sort the first four, and so on. We do it with merge sort when we sort each set of two numbers, then each set of four, then each set of eight, and so on. We do it with the Prim, Kruskal, and Dijkstra algorithms when we iterate over the nodes of a graph, and as we acquire knowledge about them, we use it to update the information about the shortest paths.

We use this approach in almost all our numerical algorithms because any differentiable function can be approximated with a linear function:

$$f(x + \delta x) \simeq f(x) + f'(x)\delta x \qquad (1.1)$$

We use this formula in the Newton method to solve nonlinear equations and optimization problems, in one or more dimensions.

We use the same approximation in the fix point method, which we use to solve equations like $f(x) = 0$; in the minimum residual and conjugate gradient methods; and to solve the Laplace equation in the last chapter of the book. In all these algorithms, we start with a random guess for the solution, and we iteratively find a better one until convergence.

The second idea of the book is that certain quantities are random, but even random numbers have patterns that we can capture using instruments like distributions and correlations. The presence of these patterns helps us model those systems that may have a random output (e.g., nuclear reactions, financial systems) and also helps us in computations. In fact, we can use random numbers to compute quantities that are not random (Monte Carlo methods). The most common approximation that we make in different parts of the book is that when a random variable x is localized at a point with a given uncertainty, δx, then its distribution is Gaussian. Thanks to the properties of Gaussian random numbers, we conclude the following:

- Using the linear approximation (our first big idea), if $z = f(x)$, the uncertainty in the output is

$$\delta z = f'(x)\delta x \tag{1.2}$$

- If we add two independent Gaussian random variables $z = x + y$, the uncertainty in the output is

$$\delta z = \sqrt{\delta x^2 + \delta y^2} \tag{1.3}$$

- If we add N independent and identically distributed Gaussian variables $z = \sum x_i$, the uncertainty in the output is

$$\delta z = \sqrt{N}\delta x \tag{1.4}$$

We use this over and over, for example, when relating the volatility over different time intervals (daily, yearly).

- If we compute an average of N independent and identically distributed Gaussian random variables, $z = 1/N \sum x_i$, the uncertainty in the average is

$$\delta z = \delta x / \sqrt{N} \tag{1.5}$$

We use this to estimate the error on the average in a Monte Carlo computation. In that case, we write it as $d\mu = \sigma/\sqrt{N}$, and σ is the standard deviation of $\{x_i\}$.

The third idea is that the time it takes to run an iterative algorithm is proportional to the number of iterations. It is therefore our goal to minimize the number of iterations required to reach a target precision. We develop a language to compare algorithms based on their running time and classify algorithms into categories. This is useful to choose the best algorithm based on the problem at hand.

In the chapter on parallel algorithms, we learn how to distribute those iterations over multiple parallel processes and how to break individual iterations into independent steps that can be executed concurrently on parallel processes, to reduce the total time required to obtain a solution within a given target precision. In the parallel case, the running time acquires an overhead that depends on the communication patterns between the parallel processes, the communication latency, and bandwidth.

In the ultimate analysis, we can even try to understand ourselves as a parallel machine that models the input from the world by approximations. The brain is a graph that can be modeled by a neural network. The learning process is an ongoing optimization process in which the brain adjusts its synapses to produce better and better responses. The decision process mimics a search tree. We solve problems by searching for the most similar problems that we have encountered before, then we refine the solution. Our DNA is a code that evolved to efficiently compress the information necessary to grow us from a single cell into a complex being. We evolved according to evolutionary mechanisms that can be modeled using genetic algorithms. We can find our similarities with other organisms using the longest common subsequence algorithm. We can reconstruct our evolutionary tree using shortest-path algorithms and find out how we came to be.

1.2 About Python

The programming language used in this book is Python [1] version 2.7. This is because Python algorithms are very similar to the corresponding pseudo-code, and therefore this language is easy to read and understand compared to other languages such as C++ or Java. Moreover, Python is a popular language in many Universities and Companies (including Google).

The goal of the book is to explain the algorithms by building them from scratch. It is not our goal to teach the user about existing libraries that may be (and often are) faster than our implementation. Two notable examples are NumPy [2] and SciPy [3]. These libraries provide a Python interface to the BLAS and LaPack libraries for linear algebra and applications. Although we wholeheartedly recommend using them when developing production code, we believe they are not appropriate for teaching the algorithms themselves because those algorithms are written in C, FORTRAN, and assembly languages and are not easy to read.

1.3 Book Structure

This book is divided into the following chapters:

- This introduction.

- An introduction to the Python programming language. The introduction assumes the reader is not new to basic programming concepts, such as conditionals, loops, and function calls, and teaches the basic syntax of the Python language, with particular focus on those built-in modules that are important for scientific applications (math, cmath, decimal, random) and a few others.

- Chapter 3 is a short review of the general theory of algorithms with applications. There we review how to determine the running time of an algorithm from simple loops to more complex recursive algorithms. We review basic data structures used to store information such as lists, arrays, stacks, queues, trees, and graphs. We also review the classification of basic algorithms such as divide-and-conquer, dynamic programming, and greedy algorithms. In the examples, we peek into complex algorithms such as Shannon–Fano compression, a maze solver, a clustering algorithm, and a neural network.

- In chapter 4, we talk about traditional numerical algorithms, in particular, linear algebra, solvers, optimizers, integrators, and Fourier–Laplace transformations. We start by reviewing the concept of Taylor series and their convergence to understand approximations, sources of error, and convergence. We then use those concepts to build more complex algorithms by systematically improving their first-order (linear) approximation. Linear algebra serves us as a tool to approximate and implement functions of many variables.

- In chapter 5, we provide a review of probability and statistics and implement basic Python functions to perform statistical analysis of random variables.

- In chapter 6, we discuss algorithms to generate random numbers from many distributions. Python already has a built-in module to generate random numbers, and in subsequent chapters, we utilize it, yet in this chapter, we discuss in detail how pseudo random number generators work and their pitfalls.

- In chapter 7, we write about Monte Carlo simulations. This is a numerical technique that utilizes random numbers to solve otherwise deterministic problems. For example, in chapter 4, we talk about numerical integration in one dimension. Those algorithms can be extended to perform numerical integration in a few (two, three, sometimes four) dimensions, but they fail for very large numbers of dimensions. That is where Monte Carlo integration comes to our rescue, as it increasingly becomes the integration method of choice as the number of variables increases. We present applications of Monte Carlo simulations.

- In chapter 8, we discuss parallel algorithms. There are many paradigms for parallel programming these days, and the tendency is toward inhomogeneous architectures. Although we review many different types of architectures, we focus on three programming paradigms that have been very successful: message-passing, map-reduce, and multi-threaded GPU programming. In the message-passing case, we create a simple "parallel simulator" (psim) in Python that allows us to understand the basic ideas behind message passing and issues with different network topologies. In the GPU case, we use pyOpenCL [4] and ocl [5], a Python-to-OpenCL compiler that allows us to write Python code and convert it in real time to OpenCL for running on the GPU.

- Finally, in the appendix, we provide a compendium of useful formulas and definitions.

1.4   Book Software

We utilize the following software libraries developed by the author and available under an Open Source BSD License:

- `http://github.com/mdipierro/nlib`

- `http://github.com/mdipierro/buckingham`

- `http://github.com/mdipierro/psim`

- `http://github.com/mdipierro/ocl`

We also utilize the following third party libraries:

- `http://www.numpy.org/`

- `http://matplotlib.org/`

- `https://github.com/michaelfairley/mincemeatpy`

- `http://mpi4py.scipy.org/`

- `http://mathema.tician.de/software/pyopencl`

All the code included in these notes is released by the author under the three-clause BSD License.

Acknowledgements

Many thanks to Alan Etkins, Brian Fox, Dan Bowker, Ethan Sudman, Holly Monteith, Konstantinos Moutselos, Michael Gheith, Paula Mikrut, Sean Neilan, and John Plamondon for reviewing different editions of this book. We also thank all the students of our classes for their useful comments and suggestions. Finally, we thank Wikipedia, from which we borrowed a few ideas and examples.

2

Overview of the Python Language

2.1 About Python

Python is a general-purpose high-level programming language. Its design philosophy emphasizes programmer productivity and code readability. It has a minimalist core syntax with very few basic commands and simple semantics. It also has a large and comprehensive standard library, including an Application Programming Interface (API) to many of the underlying operating system (OS) functions. Python provides built-in objects such as linked lists (`list`), tuples (`tuple`), hash tables (`dict`), arbitrarily long integers (`long`), complex numbers, and arbitrary precision decimal numbers.

Python supports multiple programming paradigms, including object-oriented (`class`), imperative (`def`), and functional (`lambda`) programming. Python has a dynamic type system and automatic memory management using reference counting (similar to Perl, Ruby, and Scheme).

Python was first released by Guido van Rossum in 1991 [6]. The language has an open, community-based development model managed by the nonprofit Python Software Foundation. There are many interpreters and compilers that implement the Python language, including one in Java (Jython), one built on .Net (IronPython), and one built in Python itself

(PyPy). In this brief review, we refer to the reference C implementation created by Guido.

You can find many tutorials, the official documentation, and library references of the language on the official Python website. [1]

For additional Python references, we can recommend the books in ref. [6] and ref. [7].

You may skip this chapter if you are already familiar with the Python language.

2.1.1 Python versus Java and C++ syntax

	Java/C++	Python
assignment	$a = b$;	$a = b$
comparison	if $(a == b)$	if $a == b$:
loops	for$(a = 0; a < n; a + +)$	for a in $range(0, n)$:
block	Braces {...}	indentation
function	$float f$(float a) {	def $f(a)$:
function call	$f(a)$	$f(a)$
arrays/lists	$a[i]$	$a[i]$
member	a.member	a.member
nothing	$null$ / $void*$	$None$

As in Java, variables that are primitive types (bool, int, float) are passed by copy, but more complex types, unlike C++, are passed by reference. This means when we pass an object to a function, in Python, we do not make a copy of the object, we simply define an alternate name for referencing the object in the function.

2.1.2 help, dir

The Python language provides two commands to obtain documentation about objects defined in the current scope, whether the object is built in or user defined.

We can ask for `help` about an object, for example, "1":

```
 1 >>> help(1)
 2 Help on int object:
 3
 4 class int(object)
 5  |  int(x[, base]) -> integer
 6  |
 7  |  Convert a string or number to an integer, if possible.  A floating point
 8  |  argument will be truncated towards zero (this does not include a string
 9  |  representation of a floating point number!)  When converting a string, use
10  |  the optional base.  It is an error to supply a base when converting a
11  |  non-string. If the argument is outside the integer range a long object
12  |  will be returned instead.
13  |
14  |  Methods defined here:
15  |
16  |  __abs__(...)
17  |      x.__abs__() <==> abs(x)
18 ...
```

and because "1" is an integer, we get a description about the int class and
all its methods. Here the output has been truncated because it is very
long and detailed.

Similarly, we can obtain a list of object attributes (including methods) for
any object using the command `dir`. For example:

```
 1 >>> dir(1)
 2 ['__abs__', '__add__', '__and__', '__class__', '__cmp__', '__coerce__',
 3 '__delattr__', '__div__', '__divmod__', '__doc__', '__float__',
 4 '__floordiv__', '__getattribute__', '__getnewargs__', '__hash__', '__hex__',
 5 '__index__', '__init__', '__int__', '__invert__', '__long__', '__lshift__',
 6 '__mod__', '__mul__', '__neg__', '__new__', '__nonzero__', '__oct__',
 7 '__or__', '__pos__', '__pow__', '__radd__', '__rand__', '__rdiv__',
 8 '__rdivmod__', '__reduce__', '__reduce_ex__', '__repr__', '__rfloordiv__',
 9 '__rlshift__', '__rmod__', '__rmul__', '__ror__', '__rpow__', '__rrshift__',
10 '__rshift__', '__rsub__', '__rtruediv__', '__rxor__', '__setattr__',
11 '__str__', '__sub__', '__truediv__', '__xor__']
```

2.2   Types of variables

Python is a dynamically typed language, meaning that variables do not
have a type and therefore do not have to be declared. Variables may also
change the type of value they hold through their lives. Values, on the

other hand, do have a type. You can query a variable for the type of value it contains:

```
1  >>> a = 3
2  >>> print type(a)
3  <type 'int'>
4  >>> a = 3.14
5  >>> print type(a)
6  <type 'float'>
7  >>> a = 'hello python'
8  >>> print type(a)
9  <type 'str'>
```

Python also includes, natively, data structures such as lists and dictionaries.

2.2.1 int and long

There are two types representing integer numbers: int and long. The difference is that int corresponds to the microprocessor's native bit length. Typically, this is 32 bits and can hold signed integers in range $[-2^{31}, +2^{31})$, whereas the long type can hold almost any arbitrary integer. It is important that Python automatically converts one into the other as necessary, and you can mix and match the two types in computations. Here is an example:

```
1  >>> a = 1024
2  >>> type(a)
3  <type 'int'>
4  >>> b = a**128
5  >>> print b
6  2081586438932879816385048065472817107723052449453340961063822470080721611934672 0
7  5960244788834646483696848432279085620155827671324966469298162798132113546415258 4
8  8259018778440691546366699323167100945918841095379622423387354295096957733925002 7
9  6887652058346469777062232165707683317005651120933244966378183760369413644440628 1
10 0420533968709774659160577561017394723738014294414211114063374581 76
11 >>> print type(b)
12 <type 'long'>
```

Computers represent 32-bit integer numbers by converting them to base 2. The conversion works in the following way:

```
1  def int2binary(n, nbits=32):
2      if n<0:
3          return [1 if bit==0 else 0 for bit in int2binary(-n-1,nbits)]
```

```
4    bits = [0]*nbits
5    for i in range(nbits):
6        n, bits[i] = divmod(n,2)
7    if n: raise OverflowError
8    return bits
```

The case $n < 0$ is called *two's complement* and is defined as the value obtained by subtracting the number from the largest power of 2 (2^{32} for 32 bits). Just by looking at the most significant bit, one can determine the sign of the binary number (1 for negative and 0 for zero or positive).

2.2.2   float **and** decimal

There are two ways to represent decimal numbers in Python: using the native double precision (64 bits) representation, float, or using the decimal module.

Most numerical problems are dealt with simply using float:

```
1  >>> pi = 3.141592653589793
2  >>> two_pi = 2.0 * pi
```

Floating point numbers are internally represented as follows:

$$x = \pm m 2^e \tag{2.1}$$

where x is the number, m is called the *mantissa* and is zero or a number in the range [1,2), and e is called the *exponent*. The sign, m, and e can be computed using the following algorithm, which also writes their representation in binary:

```
1  def float2binary(x,nm=4,ne=4):
2      if x==0:
3          return 0, [0]*nm, [0]*ne
4      sign,mantissa, exponent = (1 if x<0 else 0),abs(x),0
5      while abs(mantissa)>=2:
6          mantissa,exponent = 0.5*mantissa,exponent+1
7      while 0<abs(mantissa)<1:
8          mantissa,exponent = 2.0*mantissa,exponent-1
9      mantissa = int2binary(int(2**(nm-1)*mantissa),nm)
10     exponent = int2binary(exponent,ne)
11     return sign, mantissa, exponent
```

Because the exponent is stored in a fixed number of bits (11 for a 64-bit floating point number), exponents smaller than -1022 and larger than 1023 cannot be represented. An arithmetic operation that returns a number smaller than $2^{-1022} \simeq 10^{-308}$ cannot be represented and results in an underflow error. An operation that returns a number larger than $2^{1023} \simeq 10^{308}$ also cannot be represented and results in an overflow error.

Here is an example of overflow:

```
>>> a = 10.0**200
>>> a*a
inf
```

And here is an example of underflow:

```
>>> a = 10.0**-200
>>> a*a
0.0
```

Another problem with finite precision arithmetic is the loss of precision in computation. Consider the case of the difference between two numbers with very different orders of magnitude. To compute the difference, the CPU reduces them to the same exponent (the largest of the two) and then computes the difference in the two mantissas. If two numbers differ for a factor 2^k, then the mantissa of the smallest number, in binary, needs to be shifted by k positions, thus resulting in a loss of information because the k least significant bits in the mantissa are ignored. If the difference between the two numbers is greater than a factor 2^{52}, all bits in the mantissa of the smallest number are ignored, and the smallest number becomes completely invisible.

Following is a practical example that produces an incorrect result:

```
>>> a = 1.0
>>> b = 2.0**53
>>> a+b-b
0.0
```

a simple example of what occurs internally in a processor to add two floating point numbers together. The IEEE 754 standard states that for 32-bit floating point numbers, the exponent has a range of -126 to $+127$:

```
262 in IEEE 754: 0 10000111 00000110000000000000000   (+ e:8 m:1.0234375)
```

```
  3 in IEEE 754: 0 10000000 10000000000000000000000  (+ e:1 m:1.5)
265 in IEEE 754: 0 10000111 00001001000000000000000
```

To add 262.0 to 3.0, the exponents must be the same. The exponent of the lesser number is increased to the exponent of the greater number. In this case, 3's exponent must be increased by 7. Increasing the exponent by 7 means the mantissa must be shifted seven binary digits to the right:

```
0 10000111 00000110000000000000000
0 10000111 00000011000000000000000  (The implied ``1'' is also pushed seven
       places to the right)
------------------------------------
0 10000111 00001001000000000000000 which is the IEEE 754 format for 265.0
```

In the case of two numbers in which the exponent is greater than the number of digits in the mantissa, the smaller number is shifted right off the end. The effect is a zero added to the larger number.

In some cases, only some of the bits of the smaller number's mantissa are lost if a partial addition occurs.

This precision issue is always present but not always obvious. It may consist of a small discrepancy between the true value and the computed value. This difference may increase during the computation, in particular, in iterative algorithms, and may be sizable in the result of a complex algorithm.

Python also has a module for decimal floating point arithmetic that allows decimal numbers to be represented exactly. The class Decimal incorporates a notion of significant places (unlike the hardware-based binary floating point, the decimal module has a user-alterable precision):

```
>>> from decimal import Decimal, getcontext
>>> getcontext().prec = 28 # set precision
>>> Decimal(1) / Decimal(7)
Decimal('0.1428571428571428571428571429')
```

Decimal numbers can be used almost everywhere in place of floating point number arithmetic but are slower and should be used only where arbitrary precision arithmetic is required. It does not suffer from the overflow, underflow, and precision issues described earlier:

```
>>> from decimal import Decimal
>>> a = Decimal(10.0)**300
```

```
3 >>> a*a
4 Decimal('1.000000000000000000000000000E+600')
```

2.2.3 complex

Python has native support for complex numbers. The imaginary unit is represented by the character j:

```
1 >>> c = 1+2j
2 >>> print c
3 (1+2j)
4 >>> print c.real
5 1.0
6 >>> print c.imag
7 2.0
8 >>> print abs(c)
9 2.2360679775
```

The real and imaginary parts of a complex number are stored as 64-bit floating point numbers.

Normal arithmetic operations are supported. The cmath module contains trigonometric and other functions for complex numbers. For example,

```
1 >>> phi = 1j
2 >>> import cmath
3 >>> print cmath.exp(phi)
4 (0.540302305868+0.841470984808j)
```

2.2.4 str

Python supports the use of two different types of strings: ASCII strings and Unicode strings. ASCII strings are delimited by '...', "...", "'...'", or """...""". Triple quotes delimit multiline strings. Unicode strings start with a u, followed by the string containing Unicode characters. A Unicode string can be converted into an ASCII string by choosing an encoding (e.g., UTF8):

```
1 >>> a = 'this is an ASCII string'
2 >>> b = u'This is a Unicode string'
3 >>> a = b.encode('utf8')
```

After executing these three commands, the resulting a is an ASCII string storing UTF8 encoded characters.

It is also possible to write variables into strings in various ways:

```
1 >>> print 'number is ' + str(3)
2 number is 3
3 >>> print 'number is %s' % (3)
4 number is 3
5 >>> print 'number is %(number)s' % dict(number=3)
6 number is 3
```

The final notation is more explicit and less error prone and is to be preferred.

Many Python objects, for example, numbers, can be serialized into strings using str or repr. These two commands are very similar but produce slightly different output. For example,

```
1 >>> for i in [3, 'hello']:
2 ...     print str(i), repr(i)
3 3 3
4 hello 'hello'
```

For user-defined classes, str and repr can be defined and redefined using the special operators __str__ and __repr__. These are briefly described later in this chapter. For more information on the topic, refer to the official Python documentation [8].

Another important characteristic of a Python string is that it is an iterable object, similar to a list:

```
1 >>> for i in 'hello':
2 ...     print i
3 h
4 e
5 l
6 l
7 o
```

2.2.5   list and array

The distinction between lists and arrays is usually in their implementation and in the relative difference in speed of the operations they can perform. Python defines a type called list that internally is implemented more like an array.

The main methods of Python lists are append, insert, and delete. Other useful methods include count, index, reverse, and sort:

```
1  >>> b = [1, 2, 3]
2  >>> print type(b)
3  <type 'list'>
4  >>> b.append(8)
5  >>> b.insert(2, 7) # insert 7 at index 2 (3rd element)
6  >>> del b[0]
7  >>> print b
8  [2, 7, 3, 8]
9  >>> print len(b)
10 4
11 >>> b.append(3)
12 >>> b.reverse()
13 >>> print b," 3 appears ", b.count(3), " times.  The number 7 appears at index "
      , b.index(7)
14 [3, 8, 3, 7, 2] 3 appears 2 times.  The number 7 appears at index 3
```

Lists can be sliced:

```
1  >>> a= [2, 7, 3, 8]
2  >>> print a[:3]
3  [2, 7, 3]
4  >>> print a[1:]
5  [7, 3, 8]
6  >>> print a[-2:]
7  [3, 8]
```

and concatenated/joined:

```
1  >>> a = [2, 7, 3, 8]
2  >>> a = [2, 3]
3  >>> b = [5, 6]
4  >>> print a + b
5  [2, 3, 5, 6]
```

A list is iterable; you can loop over it:

```
1  >>> a = [1, 2, 3]
2  >>> for i in a:
3  ...      print i
4  1
5  2
6  3
```

A list can also be sorted in place with the sort method:

```
1  >>> a.sort()
```

There is a very common situation for which a *list comprehension* can be used. Consider the following code:

```
1  >>> a = [1,2,3,4,5]
2  >>> b = []
3  >>> for x in a:
4  ...     if x % 2 == 0:
5  ...         b.append(x * 3)
6  >>> print b
7  [6, 12]
```

This code clearly processes a list of items, selects and modifies a subset of the input list, and creates a new result list. This code can be entirely replaced with the following list comprehension:

```
1  >>> a = [1,2,3,4,5]
2  >>> b = [x * 3 for x in a if x % 2 == 0]
3  >>> print b
4  [6, 12]
```

Python has a module called `array`. It provides an efficient array implementation. Unlike lists, array elements must all be of the same type, and the type must be either a char, short, int, long, float, or double. A type of char, short, int, or long may be either signed or unsigned. Notice these are C-types, not Python types.

```
1  >>> from array import array
2  >>> a = array('d',[1,2,3,4,5])
3  array('d',[1.0, 2.0, 3.0, 4.0, 5.0])
```

An array object can be used in the same way as a list, but its elements must all be of the same type, specified by the first argument of the constructor ("d" for double, "l" for signed long, "f" for float, and "c" for character). For a complete list of available options, refer to the official Python documentation.

Using "array" over "list" can be faster, but more important, the "array" storage is more compact for large arrays.

2.2.6   tuple

A tuple is similar to a list, but its size and elements are immutable. If a tuple element is an object, the object itself is mutable, but the reference to

the object is fixed. A tuple is defined by elements separated by a comma
and optionally delimited by round parentheses:

```
>>> a = 1, 2, 3
>>> a = (1, 2, 3)
```

The round brackets are required for a tuple of zero elements such as

```
>>> a = () # this is an empty tuple
```

A trailing comma is required for a one-element tuple but not for two or
more elements:

```
>>> a = (1)   # not a tuple
>>> a = (1,) # this is a tuple of one element
>>> b = (1,2) # this is a tuple of two elements
```

Since lists are mutable; this works:

```
>>> a = [1, 2, 3]
>>> a[1] = 5
>>> print a
[1, 5, 3]
```

the element assignment does not work for a tuple:

```
>>> a = (1, 2, 3)
>>> print a[1]
2
>>> a[1] = 5
Traceback (most recent call last):
  File "<stdin>", line 1, in <module>
TypeError: 'tuple' object does not support item assignment
```

A tuple, like a list, is an iterable object. Notice that a tuple consisting of a
single element must include a trailing comma:

```
>>> a = (1)
>>> print type(a)
<type 'int'>
>>> a = (1,)
>>> print type(a)
<type 'tuple'>
```

Tuples are very useful for efficient packing of objects because of their
immutability. The brackets are often optional. You may easily get each
element of a tuple by assigning multiple variables to a tuple at one time:

```
>>> a = (2, 3, 'hello')
>>> (x, y, z) = a
>>> print x
```

```
4  2
5  >>> print z
6  hello
7  >>> a = 'alpha', 35, 'sigma' # notice the rounded brackets are optional
8  >>> p, r, q = a
9  print r
10 35
```

2.2.7 dict

A Python dict-ionary is a hash table that maps a key object to a value
object:

```
1  >>> a = {'k':'v', 'k2':3}
2  >>> print a['k']
3  v
4  >>> print a['k2']
5  3
6  >>> 'k' in a
7  True
8  >>> 'v' in a
9  False
```

You will notice that the format to define a dictionary is the same as the
JavaScript Object Notation [JSON]. Dictionaries may be nested:

```
1  >>> a = {'x':3, 'y':54, 'z':{'a':1,'b':2}}
2  >>> print a['z']
3  {'a': 1, 'b': 2}
4  >>> print a['z']['a']
5  1
```

Keys can be of any hashable type (int, string, or any object whose class
implements the __hash__ method). Values can be of any type. Different
keys and values in the same dictionary do not have to be of the same type.
If the keys are alphanumeric characters, a dictionary can also be declared
with the alternative syntax:

```
1  >>> a = dict(k='v', h2=3)
2  >>> print a['k']
3  v
4  >>> print a
5  {'h2': 3, 'k': 'v'}
```

Useful methods are has_key, keys, values, items, and update:

```
1  >>> a = dict(k='v', k2=3)
```

```
2 >>> print a.keys()
3 ['k2', 'k']
4 >>> print a.values()
5 [3, 'v']
6 >>> a.update({'n1':'new item'})     # adding a new item
7 >>> a.update(dict(n2='newer item')) # alternate method to add a new item
8 >>> a['n3'] = 'newest item'          # another method to add a new item
9 >>> print a.items()
10 [('k2', 3), ('k', 'v'), ('n3', 'newest item'), ('n2', 'newer item'), ('n1', 'new
       item')]
```

The items method produces a list of tuples, each containing a key and its associated value.

Dictionary elements and list elements can be deleted with the command del:

```
1 >>> a = [1, 2, 3]
2 >>> del a[1]
3 >>> print a
4 [1, 3]
5 >>> a = dict(k='v', h2=3)
6 >>> del a['h2']
7 >>> print a
8 {'k': 'v'}
```

Internally, Python uses the hash operator to convert objects into integers and uses that integer to determine where to store the value. Using a key that is not hashable will cause an un-hashable type error:

```
1 >>> hash("hello world")
2 -1500746465
3 >>> k = [1,2,3]
4 >>> a = {k:'4'}
5 Traceback (most recent call last):
6   File "<stdin>", line 1, in <module>
7 TypeError: unhashable type: 'list'
```

2.2.8 set

A set is something between a list and a dictionary. It represents a non-ordered list of unique elements. Elements in a set cannot be repeated. Internally, it is implemented as a hash table, similar to a set of keys in a dictionary. A set is created using the set constructor. Its argument can be a list, a tuple, or an iterator:

```
1  >>> s = set([1,2,3,4,5,5,5,5])  # notice duplicate elements are removed
2  >>> print s
3  set([1,2,3,4,5])
4  >>> s = set((1,2,3,4,5))
5  >>> print s
6  set([1,2,3,4,5])
7  >>> s = set(i for i in range(1,6))
8  >>> print s
9  set([1, 2, 3, 4, 5])
```

Sets are not ordered lists therefore appending to the end is not applicable. Instead of append, add elements to a set using the add method:

```
1  >>> s = set()
2  >>> s.add(2)
3  >>> s.add(3)
4  >>> s.add(2)
5  >>> print s
6  set([2, 3])
```

Notice that the same element cannot be added twice (2 in the example). There is no exception or error thrown when trying to add the same element more than once.

Because sets are not ordered, the order in which you add items is not necessarily the order in which they will be returned:

```
1  >>> s = set([6,'b','beta',-3.4,'a',3,5.3])
2  >>> print (s)
3  set(['a', 3, 6, 5.3, 'beta', 'b', -3.4])
```

The set object supports normal set operations like union, intersection, and difference:

```
1   >>> a = set([1,2,3])
2   >>> b = set([2,3,4])
3   >>> c = set([2,3])
4   >>> print a.union(b)
5   set([1, 2, 3, 4])
6   >>> print a.intersection(b)
7   set([2, 3])
8   >>> print a.difference(b)
9   set([1])
10  >>> if len(c) == len(a.intersection(c)):
11  ...     print "c is a subset of a"
12  ... else:
13  ...     print "c is not a subset of a"
14  ...
15  c is a subset of a
```

To check for membership,

```
1 >>> 2 in a
2 True
```

2.3 Python control flow statements

Python uses indentation to delimit blocks of code. A block starts with a line ending with colon and continues for all lines that have a similar or higher indentation as the next line:

```
1 >>> i = 0
2 >>> while i < 3:
3 ...     print i
4 ...     i = i + 1
5 0
6 1
7 2
```

It is common to use four spaces for each level of indentation. It is a good policy not to mix tabs with spaces, which can result in (invisible) confusion.

2.3.1 for...in

In Python, you can loop over iterable objects:

```
1 >>> a = [0, 1, 'hello', 'python']
2 >>> for i in a:
3 ...     print i
4 0
5 1
6 hello
7 python
```

In the preceding example, you will notice that the loop index "i" takes on the values of each element in the list [0, 1, 'hello', 'python'] sequentially. The Python range keyword creates a list of integers automatically that may be used in a "for" loop without manually creating a long list of numbers.

```
1 >>> a = range(0,5)
2 >>> print a
3 [0, 1, 2, 3, 4]
```

```
4  >>> for i in a:
5  ...     print i
6  0
7  1
8  2
9  3
10 4
```

The parameters for range(a,b,c) are as follows: the first parameter is the starting value of the list. The second parameter is the next value if the list contains one more element. The third parameter is the increment value.

The keyword range can also be called with one parameter. It is matched to "b" with the first parameter defaulting to 0 and the third to 1:

```
1  >>> print range(5)
2  [0, 1, 2, 3, 4]
3  >>> print range(53,57)
4  [53,54,55,56]
5  >>> print range(102,200,10)
6  [102, 112, 122, 132, 142, 152, 162, 172, 182, 192]
7  >>> print range(0,-10,-1)
8  [0, -1, -2, -3, -4, -5, -6, -7, -8, -9]
```

The keyword range is very convenient for creating a list of numbers; however, as the list grows in length, the memory required to store the list also grows. A more efficient option is to use the keyword xrange, which generates an iterable range instead of the entire list of elements.

This is equivalent to the C/C++/C#/Java syntax:

```
1  for(int i=0; i<4; i=i+1) { ... }
```

Another useful command is enumerate, which counts while looping and returns a tuple consisting of (index, value):

```
1  >>> a = [0, 1, 'hello', 'python']
2  >>> for (i, j) in enumerate(a):   # the ( ) around i, j are optional
3  ...     print i, j
4  0 0
5  1 1
6  2 hello
7  3 python
```

There is also a keyword range(a, b, c) that returns a list of integers starting with the value a, incrementing by c, and ending with the last value smaller than b, where a defaults to 0 and c defaults to 1.

You can jump out of a loop using `break`:

```
>>> for i in [1, 2, 3]:
...     print i
...     break
1
```

You can jump to the next loop iteration without executing the entire code block with `continue`:

```
>>> for i in [1, 2, 3]:
...     print i
...     continue
...     print 'test'
1
2
3
```

Python also supports list comprehensions, and you can build lists using the following syntax:

```
>>> a = [i*i for i in [0, 1, 2, 3]]
>>> print a
[0, 1, 4, 9]
```

Sometimes you may need a counter to "count" the elements of a list while looping:

```
>>> a = [e*(i+1) for (i,e) in enumerate(['a','b','c','d'])]
>>> print a
['a', 'bb', 'ccc', 'dddd']
```

2.3.2 while

Comparison operators in Python follow the C/C++/Java operators of ==, !=, ..., and so on. However, Python also accepts the <> operator as not equal to and is equivalent to !=. Logical operators are `and`, `or`, and `not`.

The `while` loop in Python works much as it does in many other programming languages, by looping an indefinite number of times and testing a condition before each iteration. If the condition is `False`, the loop ends:

```
>>> i = 0
>>> while i < 10:
...     i = i + 1
>>> print i
10
```

The `for` loop was introduced earlier in this chapter.

There is no `loop...until` or `do...while` construct in Python.

2.3.3   if...elif...else

The use of conditionals in Python is intuitive:

```
1  >>> for i in range(3):
2  ...     if i == 0:
3  ...         print 'zero'
4  ...     elif i == 1:
5  ...         print 'one'
6  ...     else:
7  ...         print 'other'
8  zero
9  one
10 other
```

The `elif` means "else if." Both `elif` and `else` clauses are optional. There can be more than one `elif` but only one `else` statement. Complex conditions can be created using the `not`, `and`, and `or` logical operators:

```
1  >>> for i in range(3):
2  ...     if i == 0 or (i == 1 and i + 1 == 2):
3  ...         print '0 or 1'
```

2.3.4   try...except...else...finally

Python can throw - pardon, raise - exceptions:

```
1  >>> try:
2  ...     a = 1 / 0
3  ... except Exception, e:
4  ...     print 'oops: %s' % e
5  ... else:
6  ...     print 'no problem here'
7  ... finally:
8  ...     print 'done'
9  oops: integer division or modulo by zero
10 done
```

If an exception is raised, it is caught by the `except` clause, and the `else` clause is not executed. The `finally` clause is always executed.

There can be multiple except clauses for different possible exceptions:

```
>>> try:
...     raise SyntaxError
... except ValueError:
...     print 'value error'
... except SyntaxError:
...     print 'syntax error'
syntax error
```

The finally clause is guaranteed to be executed while the except and else are not. In the following example, the function returns within a try block. This is bad practice, but it shows that the finally will execute regardless of the reason the try block is exited:

```
>>> def f(x):
...     try:
...         r = x*x
...         return r  # bad practice
...     except:
...         print "exception occurred %s" % e
...     else:
...         print "nothing else to do"
...     finally:
...         print "Finally we get here"
...
>>> y = f(3)
Finally we get here
>>> print "result is ", y
result is  9
```

For every try, you must have either an except or a finally, while the else is optional.

Here is a list of built-in Python exceptions:

```
BaseException
 +-- SystemExit
 +-- KeyboardInterrupt
 +-- Exception
      +-- GeneratorExit
      +-- StopIteration
      +-- StandardError
      |    +-- ArithmeticError
      |    |    +-- FloatingPointError
      |    |    +-- OverflowError
      |    |    +-- ZeroDivisionError
      |    +-- AssertionError
      |    +-- AttributeError
```

```
14  |      +-- EnvironmentError
15  |      |   +-- IOError
16  |      |   +-- OSError
17  |      |       +-- WindowsError (Windows)
18  |      |       +-- VMSError (VMS)
19  |      +-- EOFError
20  |      +-- ImportError
21  |      +-- LookupError
22  |      |   +-- IndexError
23  |      |   +-- KeyError
24  |      +-- MemoryError
25  |      +-- NameError
26  |      |   +-- UnboundLocalError
27  |      +-- ReferenceError
28  |      +-- RuntimeError
29  |      |   +-- NotImplementedError
30  |      +-- SyntaxError
31  |      |   +-- IndentationError
32  |      |       +-- TabError
33  |      +-- SystemError
34  |      +-- TypeError
35  |      +-- ValueError
36  |      |   +-- UnicodeError
37  |      |       +-- UnicodeDecodeError
38  |      |       +-- UnicodeEncodeError
39  |      |       +-- UnicodeTranslateError
40  +-- Warning
41      +-- DeprecationWarning
42      +-- PendingDeprecationWarning
43      +-- RuntimeWarning
44      +-- SyntaxWarning
45      +-- UserWarning
46      +-- FutureWarning
47      +-- ImportWarning
48      +-- UnicodeWarning
```

For a detailed description of each of these, refer to the official Python documentation.

Any object can be raised as an exception, but it is good practice to raise objects that extend one of the built-in exception classes.

2.3.5   def...return

Functions are declared using def. Here is a typical Python function:

```
1 >>> def f(a, b):
2 ...     return a + b
3 >>> print f(4, 2)
4 6
```

There is no need (or way) to specify the type of an argument(s) or the return value(s). In this example, a function f is defined that can take two arguments.

Functions are the first code syntax feature described in this chapter to introduce the concept of *scope*, or *namespace*. In the preceding example, the identifiers a and b are undefined outside of the scope of function f:

```
1 >>> def f(a):
2 ...     return a + 1
3 >>> print f(1)
4 2
5 >>> print a
6 Traceback (most recent call last):
7   File "<pyshell#22>", line 1, in <module>
8     print a
9 NameError: name 'a' is not defined
```

Identifiers defined outside of the function scope are accessible within the function; observe how the identifier a is handled in the following code:

```
1 >>> a = 1
2 >>> def f(b):
3 ...     return a + b
4 >>> print f(1)
5 2
6 >>> a = 2
7 >>> print f(1) # new value of a is used
8 3
9 >>> a = 1 # reset a
10 >>> def g(b):
11 ...     a = 2 # creates a new local a
12 ...     return a + b
13 >>> print g(2)
14 4
15 >>> print a # global a is unchanged
16 1
```

If a is modified, subsequent function calls will use the new value of the global a because the function definition binds the storage location of the identifier a, not the value of a itself at the time of function declaration; however, if a is assigned-to inside function g, the global a is unaffected because the new local a hides the global value. The external-scope reference can be used in the creation of *closures*:

```
1 >>> def f(x):
2 ...     def g(y):
3 ...         return x * y
4 ...     return g
5 >>> doubler = f(2) # doubler is a new function
6 >>> tripler = f(3) # tripler is a new function
7 >>> quadrupler = f(4) # quadrupler is a new function
8 >>> print doubler(5)
9 10
10 >>> print tripler(5)
11 15
12 >>> print quadrupler(5)
13 20
```

Function f creates new functions; note that the scope of the name g is entirely internal to f. Closures are extremely powerful.

Function arguments can have default values and can return multiple results as a tuple (notice the parentheses are optional and are omitted in the example):

```
1 >>> def f(a, b=2):
2 ...     return a + b, a - b
3 >>> x, y = f(5)
4 >>> print x
5 7
6 >>> print y
7 3
```

Function arguments can be passed explicitly by name; therefore the order of arguments specified in the caller can be different than the order of arguments with which the function was defined:

```
1 >>> def f(a, b=2):
2 ...     return a + b, a - b
3 >>> x, y = f(b=5, a=2)
4 >>> print x
5 7
6 >>> print y
7 -3
```

Functions can also take a runtime-variable number of arguments. Parameters that start with * and ** must be the last two parameters. If the ** parameter is used, it must be last in the list. Extra values passed in will be placed in the *identifier parameter, whereas named values will be placed into the **identifier. Notice that when passing values into the function, the unnamed values must be before any and all named values:

```
1 >>> def f(a, b, *extra, **extraNamed):
2 ...     print "a = ", a
3 ...     print "b = ", b
4 ...     print "extra = ", extra
5 ...     print "extranamed = ", extraNamed
6 >>> f(1, 2, 5, 6, x=3, y=2, z=6)
7 a =  1
8 b =  2
9 extra =  (5, 6)
10 extranamed =  {'y': 2, 'x': 3, 'z': 6}
```

Here the first two parameters (1 and 2) are matched with the parameters a and b, while the tuple 5, 6 is placed into extra and the remaining items (which are in a dictionary format) are placed into extraNamed.

In the opposite case, a list or tuple can be passed to a function that requires individual positional arguments by unpacking them:

```
1 >>> def f(a, b):
2 ...     return a + b
3 >>> c = (1, 2)
4 >>> print f(*c)
5 3
```

and a dictionary can be unpacked to deliver keyword arguments:

```
1 >>> def f(a, b):
2 ...     return a + b
3 >>> c = {'a':1, 'b':2}
4 >>> print f(**c)
5 3
```

2.3.6 lambda

The keyword lambda provides a way to define a short unnamed function:

```
1 >>> a = lambda b: b + 2
2 >>> print a(3)
3 5
```

The expression "lambda [a]:[b]" literally reads as "a function with arguments [a] that returns [b]." The lambda expression is itself unnamed, but the function acquires a name by being assigned to identifier a. The scoping rules for def apply to lambda equally, and in fact, the preceding code, with respect to a, is identical to the function declaration using def:

```
1  >>> def a(b):
2  ...     return b + 2
3  >>> print a(3)
4  5
```

The only benefit of lambda is brevity; however, brevity can be very convenient in certain situations. Consider a function called map that applies a function to all items in a list, creating a new list:

```
1  >>> a = [1, 7, 2, 5, 4, 8]
2  >>> map(lambda x: x + 2, a)
3  [3, 9, 4, 7, 6, 10]
```

This code would have doubled in size had def been used instead of lambda. The main drawback of lambda is that (in the Python implementation) the syntax allows only for a single expression; however, for longer functions, def can be used, and the extra cost of providing a function name decreases as the length of the function grows.

Just like def, lambda can be used to *curry* functions: new functions can be created by wrapping existing functions such that the new function carries a different set of arguments:

```
1  >>> def f(a, b): return a + b
2  >>> g = lambda a: f(a, 3)
3  >>> g(2)
4  5
```

Python functions created with either def or lambda allow refactoring of existing functions in terms of a different set of arguments.

2.4 Classes

Because Python is dynamically typed, Python classes and objects may seem odd. In fact, member variables (attributes) do not need to be specifically defined when declaring a class, and different instances of the same class can have different attributes. Attributes are generally associated with the instance, not the class (except when declared as "class attributes," which is the same as "static member variables" in C++/Java).

Here is an example:

```
>>> class MyClass(object): pass
>>> myinstance = MyClass()
>>> myinstance.myvariable = 3
>>> print myinstance.myvariable
3
```

Notice that pass is a do-nothing command. In this case, it is used to define a class MyClass that contains nothing. MyClass() calls the constructor of the class (in this case, the default constructor) and returns an object, an instance of the class. The (object) in the class definition indicates that our class extends the built-in object class. This is not required, but it is good practice.

Here is a more involved class with multiple methods:

```
>>> class Complex(object):
...     z = 2
...     def __init__(self, real=0.0, imag=0.0):
...         self.real, self.imag = real, imag
...     def magnitude(self):
...         return (self.real**2 + self.imag**2)**0.5
...     def __add__(self,other):
...         return Complex(self.real+other.real,self.imag+other.imag)
>>> a = Complex(1,3)
>>> b = Complex(2,1)
>>> c = a + b
>>> print c.magnitude()
5
```

Functions declared inside the class are methods. Some methods have special reserved names. For example, __init__ is the constructor. In the example, we created a class to store the real and the imag part of a complex number. The constructor takes these two variables and stores them into self (not a keyword but a variable that plays the same role as this in Java and (*this) in C++; this syntax is necessary to avoid ambiguity when declaring nested classes, such as a class that is local to a method inside another class, something Python allows but Java and C++ do not).

The self variable is defined by the first argument of each method. They all must have it, but they can use another variable name. Even if we use another name, the first argument of a method always refers to the object calling the method. It plays the same role as the this keyword in Java and C++.

Method __add__ is also a special method (all special methods start and end in double underscore) and it overloads the + operator between self and other. In the example, a+b is equivalent to a call to a.__add__(b), and the __add__ method receives self=a and other=b.

All variables are local variables of the method, except variables declared outside methods, which are called *class variables*, equivalent to C++ *static member variables*, which hold the same value for all instances of the class.

2.4.1   Special methods and operator overloading

Class attributes, methods, and operators starting with a double underscore are usually intended to be private (e.g., to be used internally but not exposed outside the class), although this is a convention that is not enforced by the interpreter.

Some of them are reserved keywords and have a special meaning:

- __len__

- __getitem__

- __setitem__

They can be used, for example, to create a container object that acts like a list:

```
1 >>> class MyList(object):
2 >>>     def __init__(self, *a): self.a = list(a)
3 >>>     def __len__(self): return len(self.a)
4 >>>     def __getitem__(self, key): return self.a[key]
5 >>>     def __setitem__(self, key, value): self.a[key] = value
6 >>> b = MyList(3, 4, 5)
7 >>> print b[1]
8 4
9 >>> b.a[1] = 7
10 >>> print b.a
11 [3, 7, 5]
```

Other special operators include __getattr__ and __setattr__, which define the get and set methods (getters and setters) for the class, and __add__, __sub__, __mul__, and __div__, which overload arithmetic operators. For the use of these operators, we refer the reader to the chapter on linear algebra, where they will be used to implement algebra for matrices.

2.4.2 class Financial Transaction

As one more example of a class, we implement a class that represents a financial transaction. We can think of a simple transaction as a single money transfer of quantity a that occurs at a given time t. We adopt the convention that a positive amount represents money flowing in and a negative value represents money flowing out.

The present value (computed at time t_0) for a transaction occurring at time t days from now of amount A is defined as

$$\mathrm{PV}(t, A) = Ae^{-tr} \tag{2.2}$$

where r is the daily risk-free interest rate. If t is measured in days, r has to be the daily risk-free return. Here we will assume it defaults to $r = 005/365$ (5% annually).

Here is a possible implementation of the transaction:

```
from datetime import date
from math import exp
today = date.today()
r_free = 0.05/365.0

class FinancialTransaction(object):
    def __init__(self,t,a,description=''):
        self.t= t
        self.a = a
        self.description = description
    def pv(self, t0=today, r=r_free):
        return self.a*exp(r*(t0-self.t).days)
    def __str__(self):
        return '%.2f dollars in %i days (%s)' % \
            (self.a, self.t, self.description)
```

Here we assume t and t_0 are datetime.date objects that store a date. The date constructor takes the year, the month, and the day separated by a comma. The expression (t0-t).days computes the distance in days between t_0 and t.

Similarly, we can implement a *Cash Flow* class to store a list of transactions, with the add method to add a new transaction to the list. The present value of a cash flow is the sum of the present values of each transaction:

```
class CashFlow(object):
    def __init__(self):
        self.transactions = []
    def add(self,transaction):
        self.transactions.append(transaction)
    def pv(self, t0, r=r_free):
        return sum(x.pv(t0,r) for x in self.transactions)
    def __str__(self):
        return '\n'.join(str(x) for x in self.transactions)
```

What is the net present value at the beginning of 2012 for a bond that pays $1000 the 20th of each month for the following 24 months (assuming a fixed interest rate of 5% per year)?

```
>>> bond = CashFlow()
>>> today = date(2012,1,1)
>>> for year in range(2012,2014):
...     for month in range(1,13):
...         coupon = FinancialTransaction(date(year,month,20), 1000)
...         bond.add(coupon)
>>> print round(bond.pv(today,r=0.05/365),0)
22826
```

This means the cost for this bond should be $22,826.

2.5 File input/output

In Python, you can open and write in a file with

```
>>> file = open('myfile.txt', 'w')
>>> file.write('hello world')
>>> file.close()
```

Similarly, you can read back from the file with

```
>>> file = open('myfile.txt', 'r')
>>> print file.read()
hello world
```

Alternatively, you can read in binary mode with "rb," write in binary mode with "wb," and open the file in append mode "a" using standard C notation.

The read command takes an optional argument, which is the number of bytes. You can also jump to any location in a file using seek :

You can read back from the file with read:

```
>>> print file.seek(6)
>>> print file.read()
world
```

and you can close the file with:

```
>>> file.close()
```

2.6 How to import modules

The real power of Python is in its library modules. They provide a large and consistent set of application programming interfaces (APIs) to many system libraries (often in a way independent of the operating system).

For example, if you need to use a random number generator, you can do the following:

```
>>> import random
>>> print random.randint(0, 9)
5
```

This prints a random integer in the range of (0,9], 5 in the example. The function randint is defined in the module random. It is also possible to import an object from a module into the current namespace:

```
>>> from random import randint
>>> print randint(0, 9)
```

or import all objects from a module into the current namespace:

```
>>> from random import *
>>> print randint(0, 9)
```

or import everything in a newly defined namespace:

```
>>> import random as myrand
>>> print myrand.randint(0, 9)
```

In the rest of this book, we will mainly use objects defined in modules math, cmath, os, sys, datetime, time, and cPickle. We will also use the random module, but we will describe it in a later chapter.

In the following subsections, we consider those modules that are most useful.

2.6.1 math **and** cmath

Here is a sampling of some of the methods available in the math and cmath packages:

- math.isinf(x) returns true if the floating point number x is positive or negative infinity

- math.isnan(x) returns true if the floating point number x is NaN; see Python documentation or IEEE 754 standards for more information

- math.exp(x) returns e**x

- math.log(x[, base] returns the logarithm of x to the optional base; if base is not supplied, e is assumed

- math.cos(x),math.sin(x),math.tan(x) returns the cos, sin, tan of the value of x; x is in radians

- math.pi, math.e are the constants for pi and e to available precision

- math.isinf(x) can be used to check if a number is *infinity*.

2.6.2 os

This module provides an interface for the operating system API:

```
1  >>> import os
2  >>> os.chdir('..')
3  >>> os.unlink('filename_to_be_deleted')
```

Some of the os functions, such as chdir, are not thread safe, for example, they should not be used in a multithreaded environment.

os.path.join is very useful; it allows the concatenation of paths in an OS-independent way:

```
1 >>> import os
2 >>> a = os.path.join('path', 'sub_path')
3 >>> print a
4 path/sub_path
```

System environment variables can be accessed via

```
1 >>> print os.environ
```

which is a read-only dictionary.

2.6.3 sys

The sys module contains many variables and functions, but used the most is sys.path. It contains a list of paths where Python searches for modules. When we try to import a module, Python searches the folders listed in sys.path. If you install additional modules in some location and want Python to find them, you need to append the path to that location to sys.path:

```
1 >>> import sys
2 >>> sys.path.append('path/to/my/modules')
```

2.6.4 datetime

The use of the datetime module is best illustrated by some examples:

```
1 >>> import datetime
2 >>> print datetime.datetime.today()
3 2008-07-04 14:03:90
4 >>> print datetime.date.today()
5 2008-07-04
```

Occasionally you may need to time stamp data based on the UTC time as opposed to local time. In this case, you can use the following function:

```
1 >>> import datetime
2 >>> print datetime.datetime.utcnow()
3 2008-07-04 14:03:90
```

The datetime module contains various classes: date, datetime, time, and timedelta. The difference between two dates or two datetimes or two time objects is a `timedelta`:

```
>>> a = datetime.datetime(2008, 1, 1, 20, 30)
>>> b = datetime.datetime(2008, 1, 2, 20, 30)
>>> c = b - a
>>> print c.days
1
```

We can also parse dates and datetimes from strings:

```
>>> s = '2011-12-31'
>>> a = datetime.datetime.strptime(s,'%Y-%m-%d')   #modified
>>> print s.year, s.day, s.month
2011 31 12   #modified
```

Notice that "%Y" matches the four-digit year, "%m" matches the month as a number (1–12), "%d" matches the day (1–31), "%H" matches the hour, "%M" matches the minute, and "%S" matches the seconds. Check the Python documentation for more options.

2.6.5 `time`

The time module differs from `date` and `datetime` because it represents time as seconds from the epoch (beginning of 1970):

```
>>> import time
>>> t = time.time()
1215138737.571
```

Refer to the Python documentation for conversion functions between time in seconds and time as a `datetime`.

2.6.6 `urllib` **and** `json`

The `urllib` is a module to download data or a web page from a URL:

```
>>> import urllib
>>> page = urllib.urlopen('http://www.google.com/')
>>> html = page.read()
```

Usually `urllib` is used to download data posted online. The challenge may be parsing the data (converting from the representation used to post it to a proper Python representation).

In the following, we create a simple helper class that can download data from Yahoo! Finance and convert each stock's historical data into a list of dictionaries. Each list element corresponds to a trading day of history of the stock, and each dictionary stores the data relative to that trading day (date, open, close, volume, adjusted close, arithmetic_return, log_return, etc.):

Listing 2.1: in file: `nlib.py`

```python
class YStock:
    """
    Class that downloads and stores data from Yahoo Finance
    Examples:
    >>> google = YStock('GOOG')
    >>> current = google.current()
    >>> price = current['price']
    >>> market_cap = current['market_cap']
    >>> h = google.historical()
    >>> last_adjusted_close = h[-1]['adjusted_close']
    >>> last_log_return = h[-1]['log_return']
    """
    URL_CURRENT = 'http://finance.yahoo.com/d/quotes.csv?s=%(symbol)s&f=%(
        columns)s'
    URL_HISTORICAL = 'http://ichart.yahoo.com/table.csv?s=%(s)s&a=%(a)s&b=%(b)s&
        c=%(c)s&d=%(d)s&e=%(e)s&f=%(f)s'
    def __init__(self,symbol):
        self.symbol = symbol.upper()

    def current(self):
        import urllib
        FIELDS = (('price', 'l1'),
                  ('change', 'c1'),
                  ('volume', 'v'),
                  ('average_daily_volume', 'a2'),
                  ('stock_exchange', 'x'),
                  ('market_cap', 'j1'),
                  ('book_value', 'b4'),
                  ('ebitda', 'j4'),
                  ('dividend_per_share', 'd'),
                  ('dividend_yield', 'y'),
                  ('earnings_per_share', 'e'),
                  ('52_week_high', 'k'),
                  ('52_week_low', 'j'),
                  ('50_days_moving_average', 'm3'),
                  ('200_days_moving_average', 'm4'),
                  ('price_earnings_ratio', 'r'),
                  ('price_earnings_growth_ratio', 'r5'),
```

```
37          ('price_sales_ratio', 'p5'),
38          ('price_book_ratio', 'p6'),
39          ('short_ratio', 's7'))
40      columns = ''.join([row[1] for row in FIELDS])
41      url = self.URL_CURRENT % dict(symbol=self.symbol, columns=columns)
42      raw_data = urllib.urlopen(url).read().strip().strip('"').split(',')
43      current = dict()
44      for i,row in enumerate(FIELDS):
45          try:
46              current[row[0]] = float(raw_data[i])
47          except:
48              current[row[0]] = raw_data[i]
49      return current
50
51  def historical(self,start=None, stop=None):
52      import datetime, time, urllib, math
53      start = start or datetime.date(1900,1,1)
54      stop = stop or datetime.date.today()
55      url = self.URL_HISTORICAL % dict(
56          s=self.symbol,
57          a=start.month-1,b=start.day,c=start.year,
58          d=stop.month-1,e=stop.day,f=stop.year)
59      # Date,Open,High,Low,Close,Volume,Adj Close
60      lines = urllib.urlopen(url).readlines()
61      raw_data = [row.split(',') for row in lines[1:] if row.count(',')==6]
62      previous_adjusted_close = 0
63      series = []
64      raw_data.reverse()
65      for row in raw_data:
66          open, high, low = float(row[1]), float(row[2]), float(row[3])
67          close, vol = float(row[4]), float(row[5])
68          adjusted_close = float(row[6])
69          adjustment = adjusted_close/close
70          if previous_adjusted_close:
71              arithmetic_return = adjusted_close/previous_adjusted_close-1.0
72
73              log_return = math.log(adjusted_close/previous_adjusted_close)
74          else:
75              arithmetic_return = log_return = None
76          previous_adjusted_close = adjusted_close
77          series.append(dict(
78              date = datetime.datetime.strptime(row[0],'%Y-%m-%d'),
79              open = open,
80              high = high,
81              low = low,
82              close = close,
83              volume = vol,
84              adjusted_close = adjusted_close,
85              adjusted_open = open*adjustment,
```

```
86          adjusted_high = high*adjustment,
87          adjusted_low = low*adjustment,
88          adjusted_vol = vol/adjustment,
89          arithmetic_return = arithmetic_return,
90          log_return = log_return))
91      return series
92
93  @staticmethod
94  def download(symbol='goog',what='adjusted_close',start=None,stop=None):
95      return [d[what] for d in YStock(symbol).historical(start,stop)]
```

Many web services return data in JSON format. JSON is slowly replacing XML as a favorite protocol for data transfer on the web. It is lighter, simpler to use, and more human readable. JSON can be thought of as serialized JavaScript. the JSON data can be converted to a Python object using a library called json:

```
1 >>> import json
2 >>> a = [1,2,3]
3 >>> b = json.dumps(a)
4 >>> print type(b)
5 <type 'str'>
6 >>> c = json.loads(b)
7 >>> a == c
8 True
```

The module json has loads and dumps methods which work very much as cPickle's methods, but they serialize the objects into a string using JSON instead of the pickle protocol.

2.6.7 pickle

This is a very powerful module. It provides functions that can serialize almost any Python object, including self-referential objects. For example, let's build a weird object:

```
1 >>> class MyClass(object): pass
2 >>> myinstance = MyClass()
3 >>> myinstance.x = 'something'
4 >>> a = [1 ,2, {'hello':'world'}, [3, 4, [myinstance]]]
```

and now:

```
1 >>> import cPickle as pickle
2 >>> b = pickle.dumps(a)
```

```
1  >>> c = pickle.loads(b)
```

In this example, b is a string representation of a, and c is a copy of a generated by deserializing b. The module pickle can also serialize to and deserialize from a file:

```
1  >>> pickle.dump(a, open('myfile.pickle', 'wb'))
2  >>> c = pickle.load(open('myfile.pickle', 'rb'))
```

2.6.8 sqlite

The Python dictionary type is very useful, but it lacks persistence because it is stored in RAM (it is lost if a program ends) and cannot be shared by more than one process running concurrently. Moreover, it is not transaction safe. This means that it is not possible to group operations together so that they succeed or fail as one.

Think for example of using the dictionary to store a bank account. The key is the account number and the value is a list of transactions. We want the dictionary to be safely stored on file. We want it to be accessible by multiple processes and applications. We want transaction safety: it should not be possible for an application to fail during a money transfer, resulting in the disappearance of money.

Python provides a module called shelve with the same interface as dict, which is stored on disk instead of in RAM. One problem with this module is that the file is not locked when accessed. If two processes try to access it concurrently, the data become corrupted. This module also does not provide transactional safety.

The proper alternative consists of using a database. There are two types of databases: relational databases (which normally use SQL syntax) and non-relational databases (often referred to as NoSQL). Key-value persistent storage databases usually follow under the latter category. Relational databases excel at storing structured data (in the form of tables), establishing relations between rows of those tables, and searches involving multiple tables linked by references. NoSQL databases excel at storing and retrieving schemaless data and replication of data (redundancy for fail safety).

Python comes with an embedded SQL database called SQLite [9]. All data in the database are stored in one single file. It supports the SQL query language and transactional safety. It is very fast and allows concurrent read (from multiple processes), although not concurrent write (the file is locked when a process is writing to the file until the transaction is committed). Concurrent write requests are queued and executed in order when the database is unlocked.

Installing and using any of these database systems is beyond the scope of this book and not necessary for our purposes. In particular, we are not concerned with relations, data replications, and speed.

As an exercise, we are going to implement a new Python class called PersistentDictionary that exposes an interface similar to a dict but uses the SQLite database for storage. The database file is created if it does not exist. PersistentDictionary will use a single table (also called persistence) to store rows containing a key (pkey) and a value (pvalue).

For later convenience, we will also add a method that can generate a UUID key. A UUID is a random string that is long enough to be, most likely, unique. This means that two calls to the same function will return different values, and the probability that the two values will be the same is negligible. Python includes a library to generate UUID strings based on a common industry standard. We use the function uuid4, which also uses the time and the IP of the machine to generate the UUID. This means the UUID is unlikely to have conflicts with (be equal to) another UUID generated on other machines. The uuid method will be useful to generate random unique keys.

We will also add a method that allows us to search for keys in the database using GLOB patterns (in a GLOB pattern, "*" represents a generic wildcard and "?" is a single-character wildcard).

Here is the code:

Listing 2.2: in file: nlib.py

```
import os
```

```
2  import uuid
3  import sqlite3
4  import cPickle as pickle
5  import unittest
6
7  class PersistentDictionary(object):
8      """
9      A sqlite based key,value storage.
10     The value can be any pickleable object.
11     Similar interface to Python dict
12     Supports the GLOB syntax in methods keys(),items(), __delitem__()
13
14     Usage Example:
15     >>> p = PersistentDictionary(path='test.sqlite')
16     >>> key = 'test/' + p.uuid()
17     >>> p[key] = {'a': 1, 'b': 2}
18     >>> print p[key]
19     {'a': 1, 'b': 2}
20     >>> print len(p.keys('test/*'))
21     1
22     >>> del p[key]
23     """
24
25     CREATE_TABLE = "CREATE TABLE persistence (pkey, pvalue)"
26     SELECT_KEYS = "SELECT pkey FROM persistence WHERE pkey GLOB ?"
27     SELECT_VALUE = "SELECT pvalue FROM persistence WHERE pkey GLOB ?"
28     INSERT_KEY_VALUE = "INSERT INTO persistence(pkey, pvalue) VALUES (?,?)"
29     UPDATE_KEY_VALUE = "UPDATE persistence SET pvalue = ? WHERE pkey = ?"
30     DELETE_KEY_VALUE = "DELETE FROM persistence WHERE pkey LIKE ?"
31     SELECT_KEY_VALUE = "SELECT pkey,pvalue FROM persistence WHERE pkey GLOB ?"
32
33     def __init__(self,
34                  path='persistence.sqlite',
35                  autocommit=True,
36                  serializer=pickle):
37         self.path = path
38         self.autocommit = autocommit
39         self.serializer = serializer
40         create_table = not os.path.exists(path)
41         self.connection  = sqlite3.connect(path)
42         self.connection.text_factory = str # do not use unicode
43         self.cursor = self.connection.cursor()
44         if create_table:
45             self.cursor.execute(self.CREATE_TABLE)
46             self.connection.commit()
47
48     def uuid(self):
49         return str(uuid.uuid4())
50
```

```
51    def keys(self,pattern='*'):
52        "returns a list of keys filtered by a pattern, * is the wildcard"
53        self.cursor.execute(self.SELECT_KEYS,(pattern,))
54        return [row[0] for row in self.cursor.fetchall()]
55
56    def __contains__(self,key):
57        return True if self.get(key)!=None else False
58
59    def __iter__(self):
60        for key in self:
61            yield key
62
63    def __setitem__(self,key, value):
64        if key in self:
65            if value is None:
66                del self[key]
67            else:
68                svalue = self.serializer.dumps(value)
69                self.cursor.execute(self.UPDATE_KEY_VALUE, (svalue, key))
70        else:
71            svalue = self.serializer.dumps(value)
72            self.cursor.execute(self.INSERT_KEY_VALUE, (key, svalue))
73        if self.autocommit: self.connection.commit()
74
75    def get(self,key):
76        self.cursor.execute(self.SELECT_VALUE, (key,))
77        row = self.cursor.fetchone()
78        return self.serializer.loads(row[0]) if row else None
79
80    def __getitem__(self, key):
81        self.cursor.execute(self.SELECT_VALUE, (key,))
82        row = self.cursor.fetchone()
83        if not row: raise KeyError
84        return self.serializer.loads(row[0])
85
86    def __delitem__(self, pattern):
87        self.cursor.execute(self.DELETE_KEY_VALUE, (pattern,))
88        if self.autocommit: self.connection.commit()
89
90    def items(self,pattern='*'):
91        self.cursor.execute(self.SELECT_KEY_VALUE, (pattern,))
92        return [(row[0], self.serializer.loads(row[1])) \
93                    for row in self.cursor.fetchall()]
94
95    def dumps(self,pattern='*'):
96        self.cursor.execute(self.SELECT_KEY_VALUE, (pattern,))
97        rows = self.cursor.fetchall()
98        return self.serializer.dumps(dict((row[0], self.serializer.loads(row[1])
                )
```

```
 99                                  for row in rows))
100
101      def loads(self, raw):
102          data = self.serializer.loads(raw)
103          for key, value in data.iteritems():
104              self[key] = value
```

This code now allows us to do the following:

- Create a persistent dictionary:

```
1 >>> p = PersistentDictionary(path='storage.sqlite',autocommit=False)
```

- Store data in it:

```
1 >>> p['some/key'] = 'some value'
```

where "some/key" must be a string and "some value" can be any Python pickleable object.

- Generate a UUID to be used as the key:

```
1 >>> key = p.uuid()
2 >>> p[key] = 'some other value'
```

- Retrieve the data:

```
1 >>> data = p['some/key']
```

- Loop over keys:

```
1 >>> for key in p: print key, p[key]
```

- List all keys:

```
1 >>> keys = p.keys()
```

- List all keys matching a pattern:

```
1 >>> keys = p.keys('some/*')
```

- List all key-value pairs matching a pattern:

```
1 >>> for key,value in p.items('some/*'): print key, value
```

- Delete keys matching a pattern:

```
1 >>> del p['some/*']
```

We will now use our persistence storage to download 2011 financial data from the SP100 stocks. This will allow us to later perform various analysis tasks on these stocks:

Listing 2.3: in file: nlib.py

```
1  >>> SP100 = ['AA', 'AAPL', 'ABT', 'AEP', 'ALL', 'AMGN', 'AMZN', 'AVP',
2  ... 'AXP', 'BA', 'BAC', 'BAX', 'BHI', 'BK', 'BMY', 'BRK.B', 'CAT', 'C', 'CL',
3  ... 'CMCSA', 'COF', 'COP', 'COST', 'CPB', 'CSCO', 'CVS', 'CVX', 'DD', 'DELL',
4  ... 'DIS', 'DOW', 'DVN', 'EMC', 'ETR', 'EXC', 'F', 'FCX', 'FDX', 'GD', 'GE',
5  ... 'GILD', 'GOOG', 'GS', 'HAL', 'HD', 'HNZ', 'HON', 'HPQ', 'IBM', 'INTC',
6  ... 'JNJ', 'JPM', 'KFT', 'KO', 'LMT', 'LOW', 'MA', 'MCD', 'MDT', 'MET',
7  ... 'MMM', 'MO', 'MON', 'MRK', 'MS', 'MSFT', 'NKE', 'NOV', 'NSC', 'NWSA',
8  ... 'NYX', 'ORCL', 'OXY', 'PEP', 'PFE', 'PG', 'PM', 'QCOM', 'RF', 'RTN', 'S',
9  ... 'SLB', 'SLE', 'SO', 'T', 'TGT', 'TWX', 'TXN', 'UNH', 'UPS', 'USB',
10 ... 'UTX', 'VZ', 'WAG', 'WFC', 'WMB', 'WMT', 'WY', 'XOM', 'XRX']
11 >>> from datetime import date
12 >>> storage = PersistentDictionary('sp100.sqlite')
13 >>> for symbol in SP100:
14 ...     key = symbol+'/2011'
15 ...     if not key in storage:
16 ...         storage[key] = YStock(symbol).historical(start=date(2011,1,1),
17 ...                                          stop=date(2011,12,31))
```

Notice that while storing one item may be slower than storing an individual item in its own files, accessing the file system becomes progressively slower as the number of files increases. Storing data in a database, long term, is a winning strategy as it scales better and it is easier to search for and extract data than it is with multiple flat files. Which type of database is most appropriate depends on the type of data and the type of queries we need to perform on the data.

2.6.9 numpy

The library numpy [2] is the Python library for efficient arrays, multidimensional arrays, and their manipulation. numpy does not ship with Python and must be installed separately.

On most platforms, this is as easy as typing in the Bash Shell:

```
1  pip install numpy
```

Yet on other platforms, it can be a more lengthy process, and we leave it to the reader to find the best installation procedure.

The basic object in numpy is the ndarray (n-dimensional array). Here we make a $10 \times 4 \times 3$ array of 64 bits float:

```
1  >>> import numpy
2  >>> a = numpy.ndarray((10,4,3),dtype=numpy.float64)
```

The class ndarray is more efficient than Python's list. It takes much less space because their elements have a fixed given type (e.g., float64). Other popular available types are: int8, int16, int32, int64, uint8, uint16, uint32, uint64, float16, float32, float64, complex64, and complex128.

We can access elements:

```
1 >>> a[0,0,0] = 1
2 >>> print a[0,0,0]
3 1.0
```

We can query for its size:

```
1 >>> print a.shape
2 (10, 4, 3)
```

We can reshape its elements:

```
1 >>> b = a.reshape((10,12))
2 >>> print a.shape
3 (10, 12)
```

We can map one type into another

```
1 >>> c = b.astype(float32)
```

We can load and save them:

```
1 >>> numpy.save('array.np',a)
2 >>> b = numpy.load('array.np')
```

And we can perform operations on them (most operations are element-wise operations):

```
1 >>> a = numpy.array([[1,2],[3,4]]) # converts a list into a ndarray
2 >>> print a
3 [[1 2]
4  [3 4]]
5 >>> print a+1
6 [[2 3]
7  [4 5]]
8 >>> print a+a
9 [[2 4]
10  [6 8]]
11 >>> print a*2
12 [[2 4]
13  [6 8]]
14 >>> print a*a
15 [[ 1  4]
16  [ 9 16]]
17 >>> print numpy.exp(a)
```

```
18 [[  2.71828183   7.3890561 ]
19  [ 20.08553692 54.59815003]]
```

The numpy module also implements common linear algebra operations:

```
1 >>> from numpy import dot
2 >>> from numpy.linalg import inv
3 >>> print dot(a,a)
4 [[ 7 10]
5  [15 22]]
6 >>> print inv(a)
7 [[-2.   1. ]
8  [ 1.5 -0.5]]
```

These operations are particularly efficient because they are implemented on top of the BLAS and LaPack libraries.

There are many other functions in the numpy module, and you can read more about it in the official documentation.

2.6.10 matplotlib

Library matplotlib [10] is the de facto standard plotting library for Python. It is one of the best and most versatile plotting libraries available. It has two modes of operation. One mode of operation, called pylab, follows a Matlab-like syntax. The other mode follows a more Python-style syntax. Here we use the latter.

You can install matplotlib with

```
1 pip install matplotlib
```

and it requires numpy. In matplotlib, we need to distinguish the following objects:

- Figure: a blank grid that can contain pairs of XY axes

- Axes: a pair of XY axes that may contain multiple superimposed plots

- FigureCanvas: a binary representation of a figure with everything that it contains

- plot: a representation of a data set such as a line plot or a scatter plot

In matplotlib, a canvas can be visualized in a window or serialized into an image file. Here we take the latter approach and create two helper functions that take data and configuration parameters and output PNG images.

We start by importing `matplotlib` and other required libraries:

Listing 2.4: in file: `nlib.py`

```
import math
import cmath
import random
import os
import tempfile
os.environ['MPLCONfigureDIR'] = tempfile.mkdtemp()
```

Now we define a helper that can plot lines, points with error bars, histograms, and scatter plots on a single canvas:

Listing 2.5: in file: `nlib.py`

```
from cStringIO import StringIO
try:
    from matplotlib.figure import Figure
    from matplotlib.backends.backend_agg import FigureCanvasAgg
    from matplotlib.patches import Ellipse
    HAVE_MATPLOTLIB = True
except ImportError:
    HAVE_MATPLOTLIB = False

class Canvas(object):

    def __init__(self, title='', xlab='x', ylab='y', xrange=None, yrange=None):
        self.fig = Figure()
        self.fig.set_facecolor('white')
        self.ax = self.fig.add_subplot(111)
        self.ax.set_title(title)
        self.ax.set_xlabel(xlab)
        self.ax.set_ylabel(ylab)
        if xrange:
            self.ax.set_xlim(xrange)
        if yrange:
            self.ax.set_ylim(yrange)
        self.legend = []

    def save(self, filename='plot.png'):
        if self.legend:
            legend = self.ax.legend([e[0] for e in self.legend],
                                    [e[1] for e in self.legend])
            legend.get_frame().set_alpha(0.7)
```

```
30      if filename:
31          FigureCanvasAgg(self.fig).print_png(open(filename, 'wb'))
32      else:
33          s = StringIO()
34          FigureCanvasAgg(self.fig).print_png(s)
35          return s.getvalue()
36
37  def binary(self):
38      return self.save(None)
39
40  def hist(self, data, bins=20, color='blue', legend=None):
41      q = self.ax.hist(data, bins)
42      #if legend:
43      #    self.legend.append((q[0], legend))
44      return self
45
46  def plot(self, data, color='blue', style='-', width=2,
47           legend=None, xrange=None):
48      if callable(data) and xrange:
49          x = [xrange[0]+0.01*i*(xrange[1]-xrange[0]) for i in xrange(0,101)]
50          y = [data(p) for p in x]
51      elif data and isinstance(data[0],(int,float)):
52          x, y = xrange(len(data)), data
53      else:
54          x, y = [p[0] for p in data], [p[1] for p in data]
55      q = self.ax.plot(x, y, linestyle=style, linewidth=width, color=color)
56      if legend:
57          self.legend.append((q[0],legend))
58      return self
59
60  def errorbar(self, data, color='black', marker='o', width=2, legend=None):
61      x,y,dy = [p[0] for p in data], [p[1] for p in data], [p[2] for p in data
             ]
62      q = self.ax.errorbar(x, y, yerr=dy, fmt=marker, linewidth=width, color=
             color)
63      if legend:
64          self.legend.append((q[0],legend))
65      return self
66
67  def ellipses(self, data, color='blue', width=0.01, height=0.01, legend=None)
         :
68      for point in data:
69          x, y = point[:2]
70          dx = point[2] if len(point)>2 else width
71          dy = point[3] if len(point)>3 else height
72          ellipse = Ellipse(xy=(x, y), width=dx, height=dy)
73          self.ax.add_artist(ellipse)
74          ellipse.set_clip_box(self.ax.bbox)
75          ellipse.set_alpha(0.5)
```

```
76          ellipse.set_facecolor(color)
77        if legend:
78          self.legend.append((q[0],legend))
79        return self
80
81    def imshow(self, data, interpolation='bilinear'):
82        self.ax.imshow(data).set_interpolation(interpolation)
83        return self
```

Notice we only make one set of axes.

The argument 111 of figure.add_subplot(111) indicates that we want a grid of 1 × 1 axes, and we ask for the first one of them (the only one).

The linesets parameter is a list of dictionaries. Each dictionary must have a "data" key corresponding to a list of (x, y) values. Each dictionary is rendered by a line connecting the points. It can have a "label," a "color," a "style," and a "width."

The pointsets parameter is a list of dictionaries. Each dictionary must have a "data" key corresponding to a list of $(x, y, \delta y)$ values. Each dictionary is rendered by a set of circles with error bars. It can optionally have a "label," a "color," and a "marker" (symbol to replace the circle).

The histsets parameter is a list of dictionaries. Each dictionary must have a "data" key corresponding to a list of x values. Each dictionary is rendered by histogram. Each dictionary can optionally have a "label" and a "color."

The ellisets parameter is also a list of dictionaries. Each dictionary must have a "data" key corresponding to a list of $(x, y, \delta x, \delta y)$ values. Each dictionary is rendered by a set of ellipses, one per point. It can optionally have a "color."

We chose to draw all these types of plots with a single function because it is common to superimpose fitting lines to histograms, points, and scatter plots.

As an example, we can plot the adjusted closing price for AAPL:

Listing 2.6: in file: nlib.py

```
1 >>> storage = PersistentDictionary('sp100.sqlite')
2 >>> appl = storage['AAPL/2011']
```

```
3 >>> points = [(x,y['adjusted_close']) for (x,y) in enumerate(appl)]
4 >>> Canvas(title='Apple Stock (2011)',xlab='trading day',ylab='adjusted close').
     plot(points,legend='AAPL').save('images/aapl2011.png')
```

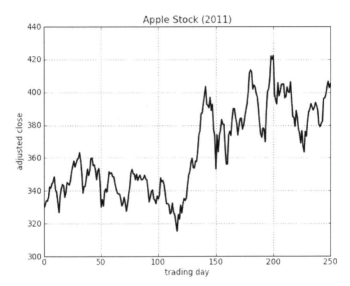

Figure 2.1: Example of a line plot. Adjusted closing price for the APPL stock in 2011 (source: Yahoo! Finance).

Here is an example of a histogram of daily arithmetic returns for the AAPL stock in 2011:

Listing 2.7: in file: nlib.py

```
1 >>> storage = PersistentDictionary('sp100.sqlite')
2 >>> appl = storage['AAPL/2011'][1:] # skip 1st day
3 >>> points = [day['arithmetic_return'] for day in appl]
4 >>> Canvas(title='Apple Stock (2011)',xlab='arithmetic return', ylab='frequency'
     ).hist(points).save('images/aapl2011hist.png')
```

Here is a scatter plot for random data points:

Listing 2.8: in file: nlib.py

```
1 >>> from random import gauss
2 >>> points = [(gauss(0,1),gauss(0,1),gauss(0,0.2),gauss(0,0.2)) for i in xrange
     (30)]
3 >>> Canvas(title='example scatter plot', xrange=(-2,2), yrange=(-2,2)).ellipses(
     points).save('images/scatter.png')
```

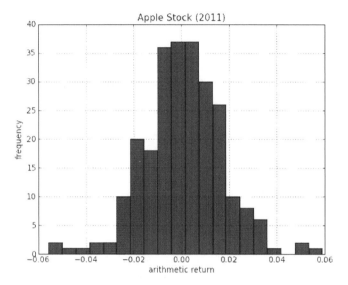

Figure 2.2: Example of a histogram plot. Distribution of daily arithmetic returns for the APPL stock in 2011 (source: Yahoo! Finance).

Here is a scatter plot showing the return and variance of the S&P100 stocks:

Listing 2.9: in file: nlib.py

```
>>> storage = PersistentDictionary('sp100.sqlite')
>>> points = []
>>> for key in storage.keys('*/2011'):
...     v = [day['log_return'] for day in storage[key][1:]]
...     ret = sum(v)/len(v)
...     var = sum(x**2 for x in v)/len(v) - ret**2
...     points.append((var*math.sqrt(len(v)),ret*len(v),0.0002,0.02))
>>> Canvas(title='S&P100 (2011)',xlab='risk',ylab='return',
...        xrange = (min(p[0] for p in points),max(p[0] for p in points)),
...        yrange = (min(p[1] for p in points),max(p[1] for p in points))
...        ).ellipses(points).save('images/sp100rr.png')
```

Notice the daily log returns have been multiplied by the number of days in one year to obtain the annual return. Similarly, the daily volatility has been multiplied by the square root of the number of days in one year to obtain the annual volatility (risk). The reason for this procedure will be explained in a later chapter.

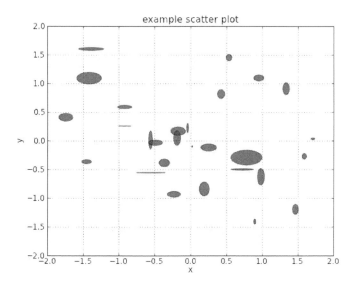

Figure 2.3: Example of a scatter plot using some random points.

Listing 2.10: in file: nlib.py

```
1 >>> def f(x,y): return (x-1)**2+(y-2)**2
2 >>> points = [[f(0.1*i-3,0.1*j-3) for i in range(61)] for j in range(61)]
3 >>> Canvas(title='example 2d function').imshow(points).save('images/color2d.png'
      )
```

The class Canvas is both in nlib.py and in the Python module canvas [11].

2.6.11   ocl

One of the best features of Python is that it can introspect itself, and this can be used to just-in-time compile Python code into other languages. For example, the Cython [12] and the ocl libraries allow decorating Python code and converting it to C code. This makes the decorated functions much faster. Cython is more powerful, and it supports a richer subset of the Python syntax; ocl instead supports only a subset of the Python syntax, which can be directly mapped into the C equivalent, but it is easier to use. Moreover, ocl can convert Python code to JavaScript and to

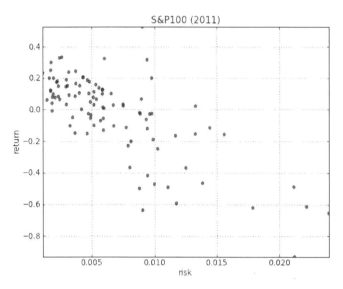

Figure 2.4: Example of a scatter plot. Risk-return plot for the S&P100 stocks in 2011 (source: Yahoo! Finance).

OpenCL (this is discussed in our last chapter).

Here is a simple example that implements the factorial function:

```
from ocl import Compiler
c99 = Compiler()

@c99.define(n='int')
def factorial(n):
    output = 1
    for k in xrange(1, n + 1):
        output = output * k
    return output
compiled = c99.compile()
print compiled.factorial(10)
assert compiled.factorial(10) == factorial(10)
```

The line @c99.define(n='int') instructs ocl that factorial must be converted to c99 and that n is an integer. The assert command checks that compiled.factorial(10) produces the same output as factorial(10), where the former runs compiled c99 code, whereas the latter runs Python code.

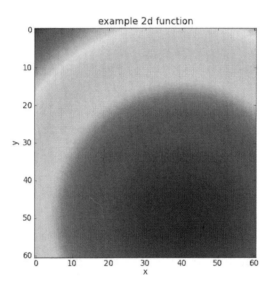

Figure 2.5: Example of a two-dimensional color plot using for $f(x, y) = (x - 1)^2 + (y - 2)^2$.

3
Theory of Algorithms

An algorithm is a step-by-step procedure for solving a problem and is typically developed before doing any programming. The word comes from *algorism*, from the mathematician al-Khwarizmi, and was used to refer to the rules of performing arithmetic using Hindu–Arabic numerals and the systematic solution of equations.

In fact, algorithms are independent of any programming language. Efficient algorithms can have a dramatic effect on our problem-solving capabilities.

The basic steps of algorithms are loops (for, conditionals (if), and function calls. Algorithms also make use of arithmetic expressions, logical expressions (not, and, or), and expressions that can be reduced to the other basic components.

The issues that concern us when developing and analyzing algorithms are the following:

1. Correctness: of the problem specification, of the proposed algorithm, and of its implementation in some programming language (we will not worry about the third one; program verification is another subject altogether)

2. Amount of work done: for example, running time of the algorithm in terms of the input size (independent of hardware and programming

language)

3. Amount of space used: here we mean the amount of extra space (system resources) beyond the size of the input (independent of hardware and programming language); we will say that an algorithm is *in place* if the amount of extra space is constant with respect to input size

4. Simplicity, clarity: unfortunately, the simplest is not always the best in other ways

5. Optimality: can we prove that it does as well as or better than any other algorithm?

3.1 Order of growth of algorithms

The *insertion sort* is a simple algorithm in which an array of elements is sorted in place, one entry at a time. It is not the fastest sorting algorithm, but it is simple and does not require extra memory other than the memory needed to store the input array.

The insertion sort works by iterating. Every iteration i of the insertion sort removes one element from the input data and inserts it into the correct position in the already-sorted subarray $A[j]$ for $0 \leq j < i$. The algorithm iterates n times (where n is the total size of the input array) until no input elements remain to be sorted:

```
def insertion_sort(A):
    for i in xrange(1,len(A)):
        for j in xrange(i,0,-1):
            if A[j]<A[j-1]:
                A[j], A[j-1] = A[j-1], A[j]
            else: break
```

Here is an example:

```
>>> import random
>>> a=[random.randint(0,100) for k in xrange(20)]
>>> insertion_sort(a)
>>> print a
[6, 8, 9, 17, 30, 31, 45, 48, 49, 56, 56, 57, 65, 66, 75, 75, 82, 89, 90, 99]
```

One important question is, how long does this algorithm take to run?

How does its running time scale with the input size?

Given any algorithm, we can define three characteristic functions:

- $T_{worst}(n)$: the running time in the worst case

- $T_{best}(n)$: the running time in the best case

- $T_{average}(n)$: the running time in the average case

The best case for an insertion sort is realized when the input is already sorted. In this case, the inner for loop exits (breaks) always at the first iteration, thus only the most outer loop is important, and this is proportional to n; therefore $T_{best}(n) \propto n$. The worst case for the insertion sort is realized when the input is sorted in reversed order. In this case, we can prove, and we do so subsequently, that $T_{worst}(n) \propto n^2$. For this algorithm, a statistical analysis shows that the worst case is also the average case.

Often we cannot determine exactly the running time function, but we may be able to set bounds to the running time.

We define the following sets:

- $O(g(n))$: the set of functions that grow no faster than $g(n)$ when $n \to \infty$

- $\Omega(g(n))$: the set of functions that grow no slower than $g(n)$ when $n \to \infty$

- $\Theta(g(n))$: the set of functions that grow at the same rate as $g(n)$ when $n \to \infty$

- $o(g(n))$: the set of functions that grow slower than $g(n)$ when $n \to \infty$

- $\omega(g(n))$: the set of functions that grow faster than $g(n)$ when $n \to \infty$

We can rewrite the preceding definitions in a more formal way:

$$O(g(n)) \equiv \{f(n) : \exists n_0, c_0, \forall n > n_0, 0 \leq f(n) < c_0 g(n)\} \quad (3.1)$$

$$\Omega(g(n)) \equiv \{f(n) : \exists n_0, c_0, \forall n > n_0, 0 \leq c_0 g(n) < f(n)\} \quad (3.2)$$

$$\Theta(g(n)) \equiv O(g(n)) \cap \Omega(g(n)) \quad (3.3)$$

$$o(g(n)) \equiv O(g(n)) - \Omega(g(n)) \quad (3.4)$$

$$w(g(n)) \equiv \Omega(g(n)) - O(g(n)) \quad (3.5)$$

We can also provide a practical rule to determine if a function f belongs to one of the previous sets defined by g.

Compute the limit

$$\lim_{n \to \infty} \frac{f(n)}{g(n)} = a \quad (3.6)$$

and look up the result in the following table:

a is positive or zero	$\implies$	$f(n) \in O(g(n)) \Leftrightarrow f \preceq g$
a is positive or infinity	$\implies$	$f(n) \in \Omega(g(n)) \Leftrightarrow f \succeq g$
a is positive	$\implies$	$f(n) \in \Theta(g(n)) \Leftrightarrow f \sim g$
a is zero	$\implies$	$f(n) \in o(g(n)) \Leftrightarrow f \prec g$
a is infinity	$\implies$	$f(n) \in w(g(n)) \Leftrightarrow f \succ g$

(3.7)

Notice the preceding practical rule assumes the limits exist.

Here is an example:

Given $f(n) = n \log n + 3n$ and $g(n) = n^2$

$$\lim_{n \to \infty} \frac{n \log n + 3n}{n^2} \xrightarrow{l'Hopital} \lim_{n \to \infty} \frac{1/n}{2} = 0 \quad (3.8)$$

we conclude that $n \log n + 3n$ is in $O(n^2)$.

Given an algorithm A that acts on input of size n, we say that the algorithm is $O(g(n))$ if its worst running time as a function of n is in $O(g(n))$. Similarly, we say that the algorithm is in $\Omega(g(n))$ if its best running time is in $\Omega(g(n))$. We also say that the algorithm is in $\Theta(g(n))$ if both its best running time and its worst running time are in $\Theta(g(n))$.

More formally, we can write the following:

$$T_{worst}(n) \in O(g(n)) \quad \Rightarrow \quad A \in O(g(n)) \qquad (3.9)$$

$$T_{best}(n) \in \Omega(g(n)) \quad \Rightarrow \quad A \in \Omega(g(n)) \qquad (3.10)$$

$$A \in O(g(n)) \text{and} A \in O(g(n)) \quad \Rightarrow \quad A \in \Theta(g(n)) \qquad (3.11)$$

$$(3.12)$$

We still have not solved the problem of computing the best, average, and worst running times.

3.1.1 Best and worst running times

The procedure for computing the worst and best running times is similar. It is simple in theory but difficult in practice because it requires an understanding of the algorithm's inner workings.

Consider the following algorithm, which finds the minimum of an array or list A:

```
def find_minimum(A):
    minimum = a[0]
    for element in A:
        if element < minimum:
            minimum = element
    return minimum
```

To compute the running time in the worst case, we assume that the maximum number of computations is performed. That happens when the if statements are always True. To compute the best running time, we assume that the minimum number of computations is performed. That happens when the if statement is always False. Under each of the two scenarios, we compute the running time by counting how many times the most nested operation is performed.

In the preceding algorithm, the most nested operation is the evaluation of the if statement, and that is executed for each element in A; for example, assuming A has n elements, the if statement will be executed n times.

Therefore both the best and worst running times are proportional to n, thus making this algorithm $O(n)$, $\Omega(n)$, and $\Theta(n)$.

More formally, we can observe that this algorithm performs the following operations:

- One assignment (line 2)
- Loops $n =$ len(A) times (line 3)
- For each loop iteration, performs one comparison (line 4)
- Line 5 is executed only if the condition is true

Because there are no nested loops, the time to execute each loop iteration is about the same, and the running time is proportional to the number of loop iterations.

For a loop iteration that does not contain further loops, the time it takes to compute each iteration, its running time, is constant (therefore equal to 1). For algorithms that contain nested loops, we will have to evaluate nested sums.

Here is the simplest example:

```
1  def loop0(n):
2      for i in xrange(0,n):
3          print i
```

which we can map into

$$T(n) = \sum_{i=0}^{i<n} 1 = n \in \Theta(n) \Rightarrow \text{loop0} \in \Theta(n) \qquad (3.13)$$

Here is a similar example where we have a single loop (corresponding to a single sum) that loops n^2 times:

```
1  def loop1(n):
2      for i in xrange(0,n*n):
3          print i
```

and here is the corresponding running time formula:

$$T(n) = \sum_{i=0}^{i<n^2} 1 = n^2 \in \Theta(n^2) \Rightarrow \texttt{loop1} \in \Theta(n^2) \qquad (3.14)$$

The following provides an example of nested loops:

```
def loop2(n):
    for i in xrange(0,n):
        for j in xrange(0,n):
            print i,j
```

Here the time for the inner loop is directly determined by n and does not depend on the outer loop's counter; therefore

$$T(n) = \sum_{i=0}^{i<n}\sum_{j=0}^{j<n} 1 = \sum_{i=0}^{i<n} n = n^2 + ... \in \Theta(n^2) \Rightarrow \texttt{loop2} \in \Theta(n^2) \qquad (3.15)$$

This is not always the case. In the following code, the inner loop does depend on the value of the outer loop:

```
def loop3(n):
    for i in xrange(0,n):
        for j in xrange(0,i):
            print i,j
```

Therefore, when we write its running time in terms of a sum, care must be taken that the upper limit of the inner sum is the upper limit of the outer sum:

$$T(n) = \sum_{i=0}^{i<n}\sum_{j=0}^{j<i} 1 = \sum_{i=0}^{i<n} i = \frac{1}{2}n(n-1) \in \Theta(n^2) \Rightarrow \texttt{loop3} \in \Theta(n^2) \qquad (3.16)$$

The appendix of this book provides examples of typical sums that come up in these types of formulas and their solutions.

Here is one more example falling in the same category, although the inner loop depends quadratically on the index of the outer loop:

Example: loop4

```python
def loop4(n):
    for i in xrange(0,n):
        for j in xrange(0,i*i):
            print i,j
```

Therefore the formula for the running time is more complicated:

$$T(n) \;=\; \sum_{i=0}^{i<n}\sum_{j=0}^{j<i^2} 1 = \sum_{i=0}^{i<n} i^2 = \frac{1}{6}n(n-1)(2n-1) \in \Theta(n^3) \quad (3.17)$$

$$\Rightarrow \quad \texttt{loop4} \in \Theta(n^3) \quad\quad\quad (3.18)$$

If the algorithm does not contain nested loops, then we need to compute the running time of each loop and take the maximum:

Example: concatenate0

```python
def concatenate0(n):
    for i in xrange(n*n):
        print i
    for j in xrange(n*n*n):
        print j
```

$$T(n) = \Theta(\max(n^2, n^3)) \Rightarrow \texttt{concatenate0} \in \Theta(n^3) \quad (3.19)$$

If there is an if statement, we need to compute the running time for each condition and pick the maximum when computing the worst running time, or the minimum for the best running time:

```python
def concatenate1(n):
    if a<0:
        for i in xrange(n*n):
            print i
    else:
        for j in xrange(n*n*n):
            print j
```

$$T_{worst}(n) = \Theta(\max(n^2, n^3)) \Rightarrow \texttt{concatenate1} \in (n^3) \quad (3.20)$$

$$T_{best}(n) = \Theta(\min(n^2, n^3)) \Rightarrow \texttt{concatenate1} \in \Omega(n^2) \quad (3.21)$$

This can be expressed more formally as follows:

$$O(f(n)) + \Theta(g(n)) = \Theta(g(n)) \text{ iff } f(n) \in O(g(n)) \qquad (3.22)$$

$$\Theta(f(n)) + \Theta(g(n)) = \Theta(g(n)) \text{ iff } f(n) \in O(g(n)) \qquad (3.23)$$

$$\Omega(f(n)) + \Theta(g(n)) = \Omega(f(n)) \text{ iff } f(n) \in \Omega(g(n)) \qquad (3.24)$$

which we can apply as in the following example:

$$T(n) = \underbrace{[n^2 + n + 3}_{\Theta(n^2)} + \underbrace{e^n - \log n]}_{\Theta(e^n)} \in \Theta(e^n) \text{ because } n^2 \in O(e^n) \qquad (3.25)$$

3.2 Recurrence relations

The *merge sort* [13] is another sorting algorithm. It is faster than the insertion sort. It was invented by John von Neumann, the physicist credited for inventing also modern computer architecture and game theory.

The merge sort works as follows.

If the input array has length 0 or 1, then it is already sorted, and the algorithm does not perform any other operation.

If the input array has a length greater than 1, it divides the array into two subsets of about half the size. Each subarray is sorted by applying the merge sort recursively (it calls itself!). It then merges the two subarrays back into one sorted array (this step is called *merge*).

Consider the following Python implementation of the merge sort:

```
 1  def mergesort(A, p=0, r=None):
 2      if r is None: r = len(A)
 3      if p<r-1:
 4          q = int((p+r)/2)
 5          mergesort(A,p,q)
 6          mergesort(A,q,r)
 7          merge(A,p,q,r)
 8
 9  def merge(A,p,q,r):
```

```
10    B,i,j = [],p,q
11    while True:
12        if A[i]<=A[j]:
13            B.append(A[i])
14            i=i+1
15        else:
16            B.append(A[j])
17            j=j+1
18        if i==q:
19            while j<r:
20                B.append(A[j])
21                j=j+1
22            break
23        if j==r:
24            while i<q:
25                B.append(A[i])
26                i=i+1
27            break
28    A[p:r]=B
```

Because this algorithm calls itself *recursively*, it is more difficult to compute its running time.

Consider the merge function first. At each step, it increases either i or j, where i is always in between p and q and j is always in between q and r. This means that the running time of the merge is proportional to the total number of values they can span from p to r. This implies that

$$\text{merge} \in \Theta(r - p) \tag{3.26}$$

We cannot compute the running time of the mergesort function using the same direct analysis, but we can assume its running time is $T(n)$, where $n = r - p$ and n is the size of the input data to be sorted and also the difference between its two arguments p and r. We can express this running time in terms of its components:

- It calls itself twice on half of the input data, $2T(n/2)$

- It calls the merge once on the entire data, $\Theta(n)$

We can summarize this into

$$T(n) = 2T(n/2) + n \tag{3.27}$$

This is called a *recurrence relation*. We turned the problem of computing the running time of the algorithm into the problem of solving the recurrence relation. This is now a math problem.

Some recurrence relations can be difficult to solve, but most of them follow in one of these categories:

$$T(n) = aT(n-b) + \Theta(f(n)) \Rightarrow T(n) \in \Theta(max(a^n, nf(n))) \tag{3.28}$$

$$T(n) = T(b) + T(n-b-a) + \Theta(f(n)) \Rightarrow T(n) \in \Theta(nf(n)) \tag{3.29}$$

$$T(n) = aT(n/b) + \Theta(n^m) \text{ and } a < b^m \Rightarrow T(n) \in \Theta(n^m) \tag{3.30}$$

$$T(n) = aT(n/b) + \Theta(n^m) \text{ and } a = b^m \Rightarrow T(n) \in \Theta(n^m \log n) \tag{3.31}$$

$$T(n) = aT(n/b) + \Theta(n^m) \text{ and } a > b^m \Rightarrow T(n) \in \Theta(n^{\log_b a}) \tag{3.32}$$

$$T(n) = aT(n/b) + \Theta(n^m \log^p n) \text{ and } a < b^m \Rightarrow T(n) \in \Theta(n^m \log^p n) \tag{3.33}$$

$$T(n) = aT(n/b) + \Theta(n^m \log^p n) \text{ and } a = b^m \Rightarrow T(n) \in \Theta(n^m \log^{p+1} n) \tag{3.34}$$

$$T(n) = aT(n/b) + \Theta(n^m \log^p n) \text{ and } a > b^m \Rightarrow T(n) \in \Theta(n^{\log_b a}) \tag{3.35}$$

$$T(n) = aT(n/b) + \Theta(q^n) \Rightarrow T(n) \in \Theta(q^n) \tag{3.36}$$

$$T(n) = aT(n/a - b) + \Theta(f(n)) \Rightarrow T(n) \in \Theta(f(n) \log(n)) \tag{3.37}$$

(they work for $m \geq 0$, $p \geq 0$, and $q > 1$).

These results are a practical simplification of a theorem known as the *master theorem* [14].

3.2.1 Reducible recurrence relations

Other recurrence relations do not immediately fit one of the preceding patterns, but often they can be reduced (transformed) to fit.

Consider the following recurrence relation:

$$T(n) = 2T(\sqrt{n}) + \log n \tag{3.38}$$

We can replace n with $e^k = n$ in eq. (3.38) and obtain

$$T(e^k) = 2T(e^{k/2}) + k \tag{3.39}$$

If we also replace $T(e^k)$ with $S(k) = T(e^k)$, we obtain

$$S(k) = 2 \underbrace{S(k/2)}_{T(e^{k/2})} + k \qquad (3.40)$$
$$\underbrace{}_{T(e^k)}$$

so that we can now apply the master theorem to S. We obtain that $S(k) \in \Theta(k \log k)$. Once we have the order of growth of S, we can determine the order of growth of $T(n)$ by substitution:

$$T(n) = S(\log n) \in \Theta(\underbrace{\log n}_{k} \log \underbrace{\log n}_{k}) \qquad (3.41)$$

Note that there are recurrence relations that cannot be solved with any of the methods described.

Here are some examples of recursive algorithms and their corresponding recurrence relations with solution:

```
def factorial1(n):
    if n==0:
        return 1
    else:
        return n*factorial1(n-1)
```

$$T(n) = T(n-1) + 1 \Rightarrow T(n) \in \Theta(n) \Rightarrow \texttt{factorial1} \in \Theta(n) \qquad (3.42)$$

```
def recursive0(n):
    if n==0:
        return 1
    else:
        loop3(n)
        return n*n*recursive0(n-1)
```

$$T(n) = T(n-1) + P_2(n) \Rightarrow T(n) \in \Theta(n^2) \Rightarrow \texttt{recursive0} \in \Theta(n^3) \qquad (3.43)$$

```
def recursive1(n):
    if n==0:
        return 1
    else:
        loop3(n)
        return n*recursive1(n-1)*recursive1(n-1)
```

$$T(n) = 2T(n-1) + P_2(n) \Rightarrow T(n) \in \Theta(2^n) \Rightarrow \texttt{recursive1} \in \Theta(2^n)$$

$$(3.44)$$

```
1  def recursive2(n):
2      if n==0:
3          return 1
4      else:
5          a=factorial0(n)
6          return a*recursive2(n/2)*recursive1(n/2)
```

$$T(n) = 2T(n/2) + P_1(n) \Rightarrow T(n) \in \Theta(n \log n) \Rightarrow \texttt{recursive2} \in \Theta(n \log n)$$

$$(3.45)$$

One example of practical interest for us is the binary search below. It finds the location of the element in a sorted input array A:

```
1  def binary_search(A,element):
2      a,b = 0, len(A)-1
3      while b>=a:
4          x = int((a+b)/2)
5          if A[x]<element:
6              a = x+1
7          elif A[x]>element:
8              b = x-1
9          else:
10             return x
11     return None
```

Notice that this algorithm does not appear to be recursive, but in practice, it is because of the apparently infinite while loop. The content of the while loop runs in constant time and then loops again on a problem of half of the original size:

$$T(n) = T(n/2) + 1 \Rightarrow \texttt{binary\_search} \in \Theta(\log n) \qquad (3.46)$$

The idea of the `binary_search` is used in the bisection method for solving nonlinear equations.

Do not confuse T notation with Θ notation:

The theta notation can also be used to describe the memory used by an

Algorithm	Recurrence Relationship	Running time
Binary Search	$T(n) = T(\frac{n}{2}) + \Theta(1)$	$\Theta(log(n))$
Binary Tree Traversal	$T(n) = 2T(\frac{n}{2}) + \Theta(1)$	$\Theta(n)$
Optimal Sorted Matrix Search	$T(n) = 2T(\frac{n}{2}) + \Theta(log(n))$	$\Theta(n)$
Merge Sort	$T(n) = T(\frac{n}{2}) + \Theta(n)$	$\Theta(nlog(n))$

algorithm as a function of the input, T_{memory}, as well as its running time.

3.3 Types of algorithms

Divide-and-conquer is a method of designing algorithms that (informally) proceeds as follows: given an instance of the problem to be solved, split this into several, smaller sub-instances (of the same problem), independently solve each of the sub-instances and then combine the sub-instance solutions to yield a solution for the original instance. This description raises the question, by what methods are the sub-instances to be independently solved? The answer to this question is central to the concept of the divide-and-conquer algorithm and is a key factor in gauging their efficiency. The solution is unique for each problem.

The merge sort algorithm of the previous section is an example of a divide-and-conquer algorithm. In the merge sort, we sort an array by dividing it into two arrays and recursively sorting (conquering) each of the smaller arrays.

Most divide-and-conquer algorithms are recursive, although this is not a requirement.

Dynamic programming is a paradigm that is most often applied in the construction of algorithms to solve a certain class of optimization problems, that is, problems that require the minimization or maximization of some measure. One disadvantage of using divide-and-conquer is that the process of recursively solving separate sub-instances can result in the same computations being performed repeatedly because identical sub-instances may arise. For example, if you are computing the path between two nodes in a graph, some portions of multiple paths will follow the same last few hops. Why compute the last few hops for every path when you would get the same result every time?

The idea behind dynamic programming is to avoid this pathology by obviating the requirement to calculate the same quantity twice. The method usually accomplishes this by maintaining a table of sub-instance results. We say that dynamic programming is a bottom-up technique in which the smallest sub-instances are explicitly solved first and the results of these are used to construct solutions to progressively larger sub-instances. In contrast, we say that the divide-and-conquer is a top-down technique.

We can refactor the mergesort algorithm to eliminate recursion in the algorithm implementation, while keeping the logic of the algorithm unchanged. Here is a possible implementation:

```
def mergesort_nonrecursive(A):
    blocksize, n = 1, len(A)
    while blocksize<n:
        for p in xrange(0, n, 2*blocksize):
            q = p+blocksize
            r = min(q+blocksize, n)
            if r>q:
                Merge(A,p,q,r)
        blocksize = 2*blocksize
```

Notice that this has the same running time as the original mergesort because, although it is not recursive, it performs the same operations:

$$T_{best} \in \Theta(n \log n) \tag{3.47}$$

$$T_{average} \in \Theta(n \log n) \tag{3.48}$$

$$T_{worst} \in \Theta(n \log n) \tag{3.49}$$

$$T_{memory} \in \Theta(1) \tag{3.50}$$

Greedy algorithms work in phases. In each phase, a decision is made that appears to be good, without regard for future consequences. Generally, this means that some local optimum is chosen. This "take what you can get now" strategy is the source of the name for this class of algorithms. When the algorithm terminates, we hope that the local optimum is equal to the global optimum. If this is the case, then the algorithm is correct; otherwise, the algorithm has produced a suboptimal solution. If the best answer is not required, then simple greedy algorithms are sometimes used to generate approximate answers, rather than using the more complicated algorithms generally required to generate an exact answer. Even for problems that can be solved exactly by a greedy algorithm, establishing the correctness of the method may be a nontrivial process.

For example, computing change for a purchase in a store is a good case of a greedy algorithm. Assume you need to give change back for a purchase. You would have three choices:

- Give the smallest denomination repeatedly until the correct amount is returned

- Give a random denomination repeatedly until you reach the correct amount. If a random choice exceeds the total, then pick another denomination until the correct amount is returned

- Give the largest denomination less than the amount to return repeatedly until the correct amount is returned

In this case, the third choice is the correct one.

Other types of algorithms do not fit into any of the preceding categories. One is, for example, backtracking. Backtracking is not covered in this course.

3.3.1 Memoization

One case of a top-down approach that is very general and falls under the umbrella of dynamic programming is called *memoization*. Memoization consists of allowing users to write algorithms using a naive divide-and-conquer approach, but functions that may be called more than once are modified so that their output is cached, and if they are called again with the same initial state, instead of the algorithm running again, the output is retrieved from the cache and returned without any computations.

Consider, for example, Fibonacci numbers:

$$\text{Fib}(0) \; = \; 0 \tag{3.51}$$
$$\text{Fib}(1) \; = \; 1 \tag{3.52}$$
$$\text{Fib}(n) \; = \; \text{Fib}(n-1) + \text{Fib}(n-2) \text{ for } n > 1 \tag{3.53}$$

which we can implement using divide-and-conquer as follows:

```
def fib(n):
    return n if n<2 else fib(n-1)+fib(n-2)
```

The recurrence relation for this algorithm is $T(n) = T(n-1) + T(n-2) + 1$, and its solution can be proven to be exponential. This is because this algorithm calls itself more than necessary with the same input values and keeps solving the same subproblem over and over.

Python can implement memoization using the following decorator:

Listing 3.1: in file: `nlib.py`

```
class memoize(object):
    def __init__ (self, f):
        self.f = f
        self.storage = {}
    def __call__ (self, *args, **kwargs):
        key = str((self.f.__name__, args, kwargs))
        try:
            value = self.storage[key]
        except KeyError:
            value = self.f(*args, **kwargs)
            self.storage[key] = value
        return value
```

and simply decorating the recursive function as follows:

Listing 3.2: in file: `nlib.py`

```
@memoize
def fib(n):
    return n if n<2 else fib(n-1)+fib(n-2)
```

which we can call as

Listing 3.3: in file: `nlib.py`

```
>>> print fib(11)
89
```

A decorator is a Python function that takes a function and returns a callable object (or a function) to replace the one passed as input. In the previous example, we are using the @memoize decorator to replace the fib function with the __call__ argument of the memoize class.

This makes the algorithm run much faster. Its running time goes from exponential to linear. Notice that the preceding memoize decorator is very general and can be used to decorate any other function.

One more direct dynamic programming approach consists in removing the recursion:

```
def fib(n):
    if n < 2: return n
    a, b = 0, 1
    for i in xrange(1,n):
        a, b = b, a+b
    return b
```

This also makes the algorithm linear and $T(n) \in \Theta(n)$.

Notice that we easily modify the memoization algorithm to store the partial results in a shared space, for example, on disk using the PersistentDictionary:

Listing 3.4: in file: nlib.py

```
class memoize_persistent(object):
    STORAGE = 'memoize.sqlite'
    def __init__ (self, f):
        self.f = f
        self.storage = PersistentDictionary(memoize_persistent.STORAGE)
    def __call__ (self, *args, **kwargs):
        key = str((self.f.__name__, args, kwargs))
        if key in self.storage:
            value = self.storage[key]
        else:
            value = self.f(*args, **kwargs)
            self.storage[key] = value
        return value
```

We can use it as we did before, but we can now start and stop the program or run concurrent parallel programs, and as long as they have access to the "memoize.sqlite" file, they will share the cache.

3.4   Timing algorithms

The order of growth is a theoretical concept. In practice, we need to time algorithms to check if findings are correct and, more important, to determine the magnitude of the constants in the T functions.

For example, consider this:

```python
def f1(n):
    return sum(g1(x) for x in range(n))

def f2(n):
    return sum(g2(x) for x in range(n**2))
```

Since f1 is $\Theta(n)$ and f2 is $\Theta(n^2)$, we may be led to conclude that the latter is slower. It may very well be that g1 is 10^6 smaller than g2 and therefore $T_{f1}(n) = c_1 n$, $T_{f2}(n) = c_2 n^2$, but if $c_1 = 10^6 c_2$, then $T_{f1}(n) > T_{f2}(n)$ when $n < 10^6$.

To time functions in Python, we can use this simple algorithm:

```python
def timef(f, ns=1000, dt = 60):
    import time
    t = t0 = time.time()
    for k in xrange(1,ns):
        f()
        t = time.time()
        if t-t0>dt: break
    return (t-t0)/k
```

This function calls and averages the running time of f() for the minimum between ns=1000 iterations and dt=60 seconds.

It is now easy, for example, to time the fib function without memoize,

```python
>>> def fib(n):
...     return n if n<2 else fib(n-1)+fib(n-2)
>>> for k in range(15,20):
...     print k,timef(lambda:fib(k))
15 0.000315684575338
16 0.000576375363706
17 0.000936052104732
18 0.00135168084153
19 0.00217730337912
```

and with memoize,

```python
>>> @memoize
```

```
2  ... def fib(n):
3  ...     return n if n<2 else fib(n-1)+fib(n-2)
4  >>> for k in range(15,20):
5  ...     print k,timef(lambda:fib(k))
6  15 4.24022311802e-06
7  16 4.02901146386e-06
8  17 4.21922128122e-06
9  18 4.02495429084e-06
10 19 3.73784963552e-06
```

The former shows an exponential behavior; the latter does not.

3.5 Data structures

3.5.1 Arrays

An array is a data structure in which a series of numbers are stored contiguously in memory. The time to access each number (to read or write it) is constant. The time to remove, append, or insert an element may require moving the entire array to a more spacious memory location, and therefore, in the worst case, the time is proportional to the size of the array.

Arrays are the appropriate containers when the number of elements does not change often and when elements have to be accessed in random order.

3.5.2 List

A list is a data structure in which data are not stored contiguously, and each element has knowledge of the location of the next element (and perhaps of the previous element, in a doubly linked list). This means that accessing any element for (read and write) requires finding the element and therefore looping. In the worst case, the time to find an element is proportional to the size of the list. Once an element has been found, any operation on the element, including read, write, delete, and insert, before or after can be done in constant time.

Lists are the appropriate choice when the number of elements can vary often and when their elements are usually accessed sequentially via iterations.

In Python, what is called a list is actually an array of pointers to the elements.

3.5.3 Stack

A stack data structure is a container, and it is usually implemented as a list. It has the property that the first thing you can take out is the last thing put in. This is commonly known as last-in, first-out, or LIFO. The method to insert or add data to the container is called *push*, and the method to extract data is called *pop*.

In Python, we can implement push by appending an item at the end of a list (Python already has a method for this called .append), and we can implement pop by removing the last element of a list and returning it (Python has a method for this called .pop).

A simple stack example is as follows:

```
1  >>> stk = []
2  >>> stk.append("One")
3  >>> stk.append("Two")
4  >>> print stk.pop()
5  Two
6  >>> stk.append("Three")
7  >>> print stk.pop()
8  Three
9  >>> print stk.pop()
10 One
```

3.5.4 Queue

A queue data structure is similar to a stack but, whereas the stack returns the most recent item added, a queue returns the oldest item in the list. This is commonly called first-in, first-out, or FIFO. To use Python lists to implement a queue, insert the element to add in the first position of the list as follows:

```
1 >>> que = []
2 >>> que.insert(0,"One")
3 >>> que.insert(0,"Two")
4 >>> print que.pop()
5 One
6 >>> que.insert(0,"Three")
7 >>> print que.pop()
8 Two
9 >>> print que.pop()
10 Three
```

Lists in Python are not an efficient mechanism for implementing queues. Each insertion or removal of an element at the front of a list requires all the elements in the list to be shifted by one. The Python package collections.deque is designed to implement queues and stacks. For a stack or queue, you use the same method .append to add items. For a stack, .pop is used to return the most recent item added, while to build a queue, use .popleft to remove the oldest item in the list:

```
1 >>> from collections import deque
2 >>> que = deque([])
3 >>> que.append("One")
4 >>> que.append("Two")
5 >>> print que.popleft()
6 One
7 >>> que.append("Three")
8 >>> print que.popleft()
9 Two
10 >>> print que.popleft()
11 Three
```

3.5.5 Sorting

In the previous sections, we have seen the *insertion sort* and the *merge sort*. Here we consider, as examples, other sorting algorithms: the *quicksort* [13], the *randomized quicksort*, and the *counting sort*:

```
1 def quicksort(A,p=0,r=-1):
2     if r is -1:
3         r=len(A)
4     if p<r-1:
5         q=partition(A,p,r)
6         quicksort(A,p,q)
7         quicksort(A,q+1,r)
8
```

```
 9  def partition(A,i,j):
10      x=A[i]
11      h=i
12      for k in xrange(i+1,j):
13          if A[k]<x:
14              h=h+1
15              A[h],A[k] = A[k],A[h]
16      A[h],A[i] = A[i],A[h]
17      return h
```

The running time of the quicksort is given by

$$T_{best} \quad \in \quad \Theta(n \log n) \tag{3.54}$$

$$T_{average} \quad \in \quad \Theta(n \log n) \tag{3.55}$$

$$T_{worst} \quad \in \quad \Theta(n^2) \tag{3.56}$$

$$\tag{3.57}$$

The quicksort can also be randomized by picking the pivot, A[r], at random:

```
1  def quicksort(A,p=0,r=-1):
2      if r is -1:
3          r=len(A)
4      if p<r-1:
5          q = random.randint(p,r-1)
6          A[p], A[q] = A[q], A[p]
7          q=partition(A,p,r)
8          quicksort(A,p,q)
9          quicksort(A,q+1,r)
```

In this case, the best and the worst running times do not change, but the average improves when the input is already almost sorted.

The *counting sort* algorithm is special because it only works for arrays of positive integers. This extra requirement allows it to run faster than other sorting algorithms, under some conditions. In fact, this algorithm is linear in the range span by the elements of the input array.

Here is a possible implementation:

```
1  def countingsort(A):
2      if min(A)<0:
3          raise '_counting_sort List Unbound'
```

```
 4    i, n, k = 0, len(A), max(A)+1
 5    C = [0]*k
 6    for j in xrange(n):
 7        C[A[j]] = C[A[j]]+1
 8    for j in xrange(k):
 9        while C[j]>0:
10            (A[i], C[j], i) = (j, C[j]-1, i+1)
```

If we define $k = max(A) - min(A) + 1$ and $n = len(A)$, we see

$$T_{best} \quad \in \quad \Theta(k+n) \tag{3.58}$$

$$T_{average} \quad \in \quad \Theta(k+n) \tag{3.59}$$

$$T_{worst} \quad \in \quad \Theta(k+n) \tag{3.60}$$

$$T_{memory} \quad \in \quad \Theta(k) \tag{3.61}$$

Notice that here we have also computed T_{memory}, for example, the order of growth of memory (not of time) as a function of the input size. In fact, this algorithm differs from the previous ones because it requires a temporary array C.

3.6 Tree algorithms

3.6.1 Heapsort and priority queues

Consider a *complete binary tree* as the one in the following figure:

It starts from the top node, called the *root*. Each node has zero, one, or two children. It is called complete because nodes have been added from top to bottom and left to right, filling available slots. We can think of each level of the tree as a generation, where the older generation consists of one node, the next generation of two, the next of four, and so on. We can also number nodes from top to bottom and left to right, as in the image. This allows us to map the elements of a complete binary tree into the elements of an array.

We can implement a complete binary tree using a list, and the child–parent relations are given by the following formulas:

```
 1  def heap_parent(i):
```

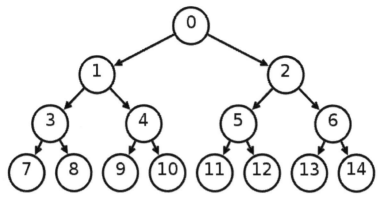

Figure 3.1: Example of a heap data structure. The number represents not the data in the heap but the numbering of the nodes.

```
2        return int((i-1)/2)
3
4    def heap_left_child(i):
5        return 2*i+1
6
7    def heap_right_child(i):
8        return 2*i+2
```

We can store data (e.g., numbers) in the nodes (or in the corresponding array). If the data are stored in such a way that the value at one node is always greater or equal than the value at its children, the array is called a *heap* and also a *priority queue*.

First of all, we need an algorithm to convert a list into a heap:

```
1    def heapify(A):
2        for i in xrange(int(len(A)/2)-1,-1,-1):
3            heapify_one(A,i)
4
5    def heapify_one(A,i,heapsize=None):
6        if heapsize is None:
7            heapsize = len(A)
8        left = 2*i+1
9        right = 2*i+2
10       if left<heapsize and A[left]>A[i]:
11           largest = left
12       else:
13           largest = i
14       if right<heapsize and A[right]>A[largest]:
15           largest = right
```

```
16    if largest!=i:
17        (A[i], A[largest]) = (A[largest], A[i])
18        heapify_one(A,largest,heapsize)
```

Now we can call build_heap on any array or list and turn it into a heap. Because the first element is by definition the smallest, we can use the heap to sort numbers in three steps:

- We turn the array into a heap

- We extract the largest element

- We apply recursion by sorting the remaining elements

Instead of using the preceding divide-and-conquer approach, it is better to use a dynamic programming approach. When we extract the largest element, we swap it with the last element of the array and make the heap one element shorter. The new, shorter heap does not need a full build_heap step because the only element out of order is the root node. We can fix this by a single call to heapify.

This is a possible implementation for the heapsort [15]:

```
1  def heapsort(A):
2      heapify(A)
3      n = len(A)
4      for i in xrange(n-1,0,-1):
5          (A[0],A[i]) = (A[i],A[0])
6          heapify_one(A,0,i)
```

In the average and worst cases, it runs as fast as the quicksort, but in the best case, it is linear:

$$T_{best} \in \Theta(n) \tag{3.62}$$

$$T_{average} \in \Theta(n \log n) \tag{3.63}$$

$$T_{worst} \in \Theta(n \log n) \tag{3.64}$$

$$T_{memory} \in \Theta(1) \tag{3.65}$$

A heap can be used to implement a priority queue, for example, storage from which we can efficiently extract the largest element.

All we need is a function that allows extracting the root element from a

heap (as we did in the heapsort and heapify of the remaining data) and a
function to push a new value into the heap:

```
1  def heap_pop(A):
2      if len(A)<1:
3          raise RuntimeError('Heap Underflow')
4      largest = A[0]
5      A[0] = A[len(A)-1]
6      del A[len(A)-1]
7      heapify_one(A,0)
8      return largest
9
10 def heap_push(A,value):
11     A.append(value)
12     i = len(A)-1
13     while i>0:
14         j = heap_parent(i)
15         if A[j]<A[i]:
16             (A[i],A[j],i) = (A[j],A[i],j)
17         else:
18             break
```

The running times for heap_pop and heap_push are the same:

$$T_{best} \in \Theta(1) \tag{3.66}$$

$$T_{average} \in \Theta(\log n) \tag{3.67}$$

$$T_{worst} \in \Theta(\log n) \tag{3.68}$$

$$T_{memory} \in \Theta(1) \tag{3.69}$$

Here is an example:

```
1  >>> a = [6,2,7,9,3]
2  >>> heap = []
3  >>> for element in a: heap_push(heap,element)
4  >>> while heap: print heap_pop(heap)
5  9
6  7
7  6
8  3
9  2
```

Heaps find application in many numerical algorithms. In fact, there is
a built-in Python module for them called heapq, which provides similar
functionality to the functions defined here, except that we defined a max

heap (pops the max element) while `heapq` is a min heap (pops the minimum):

```
1 >>> from heapq import heappop, heappush
2 >>> a = [6,2,7,9,3]
3 >>> heap = []
4 >>> for element in a: heappush(heap,element)
5 >>> while heap: print heappop(heap)
6 9
7 7
8 6
9 3
10 2
```

Notice `heappop` instead of `heap_pop` and `heappush` instead of `heap_push`.

3.6.2   Binary search trees

A binary tree is a tree in which each node has at most two children (left and right). A binary tree is called a *binary search tree* if the value of a node is always greater than or equal to the value of its left child and less than or equal to the value of its right child.

A binary search tree is a kind of storage that can efficiently be used for searching if a particular value is in the storage. In fact, if the value for which we are looking is less than the value of the root node, we only have to search the left branch of the tree, and if the value is greater, we only have to search the right branch. Using divide-and-conquer, searching each branch of the tree is even simpler than searching the entire tree because it is also a tree, but smaller.

This means that we can search simply by traversing the tree from top to bottom along some path down the tree. We choose the path by moving down and turning left or right at each node, until we find the element for which we are looking or we find the end of the tree. We can search $T(d)$, where d is the depth of the tree. We will see later that it is possible to build binary trees where $d = \log n$.

To implement it, we need to have a class to represent a binary tree:

```
1 class BinarySearchTree(object):
2     def __init__(self):
```

```
3          self.left = self.right = None
4          self.key = self.value =  None
5      def __setitem__(self,key,value):
6          if self.key == None:
7              self.key, self.value = key, value
8          elif key == self.key:
9              self.value = value
10         elif key < self.key:
11             if self.left:
12                 self.left[key] = value
13             else:
14                 self.left = BinarySearchTree(key,value)
15         else:
16             if self.right:
17                 self.right[key] = value
18             else:
19                 self.right = BinarySearchTree(key,value)
20     def __getitem__(self,key):
21         if self.key == None:
22             retur None
23         elif key == self.key:
24             return self.value
25         elif key<self.key and self.left:
26             return self.left[key]
27         elif key>self.key and self.right:
28             return self.right[key]
29         else:
30             return None
31     def min(self):
32         node = self
33         while node.left:
34             node = self.left
35         return node.key, node.value
36     def max(self):
37         node = self
38         while node.right:
39             node = self.right
40         return node.key, node.value
```

The binary tree can be used as follows:

```
1  >>> root = BinarySearchTree()
2  >>> root[5] = 'aaa'
3  >>> root[3] = 'bbb'
4  >>> root[8] = 'ccc'
5  >>> print root.left.key
6  3
7  >>> print root.left.value
8  bbb
```

```
 9 >>> print root[3]
10 bbb
11 >>> print root.max()
12 8 ccc
```

Notice that an empty tree is treated as an exception, where key = None.

3.6.3   Other types of trees

There are many other types of trees.

For example, AVL trees are binary search trees that are rebalanced after each insertion or deletion. They are rebalanced in such a way that for each node, the height of the left subtree minus the height of the right subtree is more or less the same. The rebalance operation can be done in $O(\log n)$.

For an AVL tree, the time for inserting or removing an element is given by

$$T_{best} \quad \in \quad \Theta(1) \tag{3.70}$$
$$T_{average} \quad \in \quad \Theta(\log n) \tag{3.71}$$
$$T_{worst} \quad \in \quad \Theta(\log n) \tag{3.72}$$
$$\tag{3.73}$$

Until now, we have considered binary trees (each node has two children and stores one value). We can generalize this to k trees, for which each node has k children and stores more than one value.

B-trees are a type of k-tree optimized to read and write large blocks of data. They are normally used to implement database indices and are designed to minimize the amount of data to move when the tree is rebalanced.

3.7 Graph algorithms

A *graph* G is a set of *vertices* V and a set of *links* (also called *edges*) connecting those vertices E. Each link connects one vertex to another.

As an example, you can think of a set of cities connected by roads. The cities are the vertices and the roads are the links.

A link may have attributes. In the case of a road, it could be the name of the road or its length.

In general, a link, indicated with the notation e_{ij}, connecting vertex i with vertex j is called a *directed link*. If the link has no direction $e_{ij} = e_{ji}$, it is called an undirected link. A graph that contains only undirected links is an *undirected graph*; otherwise, it is a *directed graph*.

In the road analogy, some roads can be "one way" (directed links) and some can be "two way" (undirected links).

A *walk* is an alternating sequence of vertices and links, with each link being incident to the vertices immediately preceding and succeeding it in the sequence. A *trail* is a walk with no repeated links.

A *path* is a walk with no repeated vertices. A walk is closed if the initial vertex is also the terminal vertex.

A *cycle* is a closed trail with at least one edge and with no repeated vertices, except that the initial vertex is also the terminal vertex.

A graph that contains no cycles is an *acyclic graph*. Any connected acyclic undirected graph is also a *tree*.

A *loop* is a one-link path connecting a vertex with itself.

A non null graph is *connected* if, for every pair of vertices, there is a walk whose ends are the given vertices. Let us write $i \tilde{\ } j$ if there is a path from i to j. Then $\tilde{\ }$ is an equivalence relation. The equivalence classes under $\tilde{\ }$ are the vertex sets of the connected components of G. A connected graph is therefore a graph with exactly one connected component.

A graph is called *complete* when every pair of vertices is connected by a link (or edge).

A *clique* of a graph is a subset of vertices in which every pair is an edge.

The *degree* of a vertex of a graph is the number of edges incident to it.

If i and j are vertices, the *distance* from i to j, written d_{ij}, is the minimum length of any path from i to j. In a connected undirected graph, the length of links induces a metric because for every two vertices, we can define their distance as the length of the shortest path connecting them.

The *eccentricity*, $e(i)$, of the vertex i is the maximum value of d_{ij}, where j is allowed to range over all of the vertices of the graph. This gives the largest shortest distance to any connected node in the graph.

The *subgraph* of G induced by a subset W of its vertices V ($W \subseteq V$) is the graph formed by the vertices in W and all edges whose two endpoints are in W.

The graph is the more complex of the data structures considered so far because it includes the tree as a particular case (yes, a tree is also a graph, but in general, a graph is not a tree), and the tree includes a list as a particular case (yes, a list is a tree in which every node has no more than one child); therefore a list is also a particular case of a graph.

The graph is such a general data structure that it can be used to model the brain. Think of neurons as vertices and synapses as links connecting them. We push this analogy later by implementing a simple neural network simulator.

In what follows, we represent a graph in the following way, where links are edges:

```
1 >>> vertices = ['A','B','C','D','E']
2 >>> links = [(0,1),(1,2),(1,3),(2,5),(3,4),(3,2)]
3 >>> graph = (vertices, links)
```

Vertices are stored in a list or array and so are links. Each link is a tuple containing the ID of the source vertex, the ID of the target vertex, and perhaps optional parameters. Optional parameters are discussed later, but for now, they may include link details such as length, speed, reliability, or billing rate.

3.7.1 Breadth-first search

The breadth-first search [16] (BFS) is an algorithm designed to visit all vertices in a connected graph. In the cities analogy, we are looking for a travel strategy to make sure we visit every city reachable by roads, once and only once.

The algorithm begins at one vertex, the origin, and expands out, eventually visiting each node in the graph that is somehow connected to the origin vertex. Its main feature is that it explores the neighbors of the current vertex before moving on to explore remote vertices and their neighbors. It visits other vertices in the same order in which they are discovered.

The algorithm starts by building a table of neighbors so that for each vertex, it knows which other vertices it is connected to. It then maintains two lists, a list of blacknodes (defined as vertices that have been visited) and graynodes (defined as vertices that have been discovered because the algorithm has visited its neighbor). It returns a list of blacknodes in the order in which they have been visited.

Here is the algorithm:

Listing 3.5: in file: `nlib.py`

```
def breadth_first_search(graph,start):
    vertices, link = graph
    blacknodes = []
    graynodes = [start]
    neighbors = [[] for vertex in vertices]
    for link in links:
        neighbors[link[0]].append(link[1])
    while graynodes:
        current = graynodes.pop()
        for neighbor in neighbors[current]:
            if not neighbor in blacknodes+graynodes:
                graynodes.insert(0,neighbor)
        blacknodes.append(current)
    return blacknodes
```

The BFS algorithm scales as follows:

$$T_{best} \quad \in \quad \Theta(n_E + n_V) \tag{3.74}$$

$$T_{average} \quad \in \quad \Theta(n_E + n_V) \tag{3.75}$$

$$T_{worst} \quad \in \quad \Theta(n_E + n_V) \tag{3.76}$$

$$T_{memory} \quad \in \quad \Theta(n) \tag{3.77}$$

3.7.2 Depth-first search

The depth-first search [17] (DFS) algorithm is very similar to the BFS, but it takes the opposite approach and explores as far as possible along each branch before backtracking.

In the cities analogy, if the BFS was exploring cities in the neighborhood before moving farther away, the DFS does the opposite and brings us first to distant places before visiting other nearby cities.

Here is a possible implementation:

Listing 3.6: in file: nlib.py

```
def depth_first_search(graph,start):
    vertices, link = graph
    blacknodes = []
    graynodes = [start]
    neighbors = [[] for vertex in vertices]
    for link in links:
        neighbors[link[0]].append(link[1])
    while graynodes:
        current = graynodes.pop()
        for neighbor in neighbors[current]:
            if not neighbor in blacknodes+graynodes:
                graynodes.append(neighbor)
        blacknodes.append(current)
    return blacknodes
```

Notice that the BFS and the DFS differ for a single line, which determines whether graynodes is a queue (BSF) or a stack (DFS). When graynodes is a queue, the first vertex discovered is the first visited. When it is a stack, the last vertex discovered is the first visited.

The DFS algorithm goes as follows:

$$T_{best} \;\in\; \Theta(n_E + n_V) \tag{3.78}$$

$$T_{average} \;\in\; \Theta(n_E + n_V) \tag{3.79}$$

$$T_{worst} \;\in\; \Theta(n_E + n_V) \tag{3.80}$$

$$T_{memory} \;\in\; \Theta(1) \tag{3.81}$$

3.7.3   Disjoint sets

This is a data structure that can be used to store a set of sets and implements efficiently the join operation between sets. Each element of a set is identified by a representative element. The algorithm starts by placing each element in a set of its own, so there are n initial disjoint sets. Each is represented by itself. When two sets are joined, the representative element of the latter is made to point to the representative element of the former. The set of sets is stored as an array of integers. If at position i the array stores a negative number, this number is interpreted as being the representative element of its own set. If the number stored at position i is instead a nonnegative number j, it means that it belongs to a set that was joined with the set containing j.

Here is the implementation:

Listing 3.7: in file: `nlib.py`

```
class DisjointSets(object):
    def __init__(self,n):
        self.sets = [-1]*n
        self.counter = n
    def parent(self,i):
        while True:
            j = self.sets[i]
            if j<0:
                return i
            i = j
    def join(self,i,j):
        i,j = self.parent(i),self.parent(j)
        if i!=j:
            self.sets[i] += self.sets[j]
            self.sets[j] = i
            self.counter-=1
```

```
17         return True  # they have been joined
18       return False    # they were already joined
19    def joined(self,i,j):
20       return self.parent(i) == self.parent(j)
21    def __len__(self):
22       return self.counter
```

Notice that we added a member variable counter that is initialized to the number of disjoint sets and is decreased by one every time two sets are merged. This allows us to keep track of how many disjoint sets exist at each time. We also override the __len__ operator so that we can check the value of the counter using the len function on a DisjointSet.

As an example of application, here is a code that builds a n^d maze. It may be easier to picture it with $d = 2$, a two-dimensional maze. The algorithm works by assuming there is a wall connecting any couple of two adjacent cells. It labels the cells using an integer index. It puts all the cells into a DisjointSets data structure and then keeps tearing down walls at random. Two cells on the maze belong to the same set if they are connected, for example, if there is a path that connects them. At the beginning, each cell is its own set because it is isolated by walls. Walls are torn down by being removed from the list wall if the wall was separating two disjoint sets of cells. Walls are torn down until all cells belong to the same set, for example, there is a path connecting any cell to any cell:

```
1  def make_maze(n,d):
2      walls = [(i,i+n**j) for i in xrange(n**2) for j in xrange(d) if (i/n**j)%n
           +1<n]
3      torn_down_walls = []
4      ds = DisjointSets(n**d)
5      random.shuffle(walls)
6      for i,wall in enumerate(walls):
7          if ds.join(wall[0],wall[1]):
8              torn_down_walls.append(wall)
9          if len(ds)==1:
10              break
11      walls = [wall for wall in walls if not wall in torn_down_walls]
12      return walls, torn_down_walls
```

Here is an example of how to use it. This example also draws the walls and the border of the maze:

```
1  >>> walls, torn_down_walls = make_maze(n=20,d=2)
```

The following figure shows a representation of a generated maze:

Figure 3.2: Example of a maze as generated using the DisjointSets algorithm.

3.7.4 Minimum spanning tree: Kruskal

Given a connected graph with weighted links (links with a weight or length), a *minimum spanning tree* is a subset of that graph that connects all vertices of the original graph, and the sum of the link weights is minimal. This subgraph is also a tree because the condition of minimal weight implies that there is only one path connecting each couple of vertices.

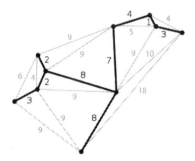

Figure 3.3: Example of a minimum spanning tree subgraph of a larger graph. The numbers on the links indicate their weight or length.

One algorithm to build the minimal spanning tree of a graph is the Kruskal [18] algorithm. It works by placing all vertices in a DisjointSets structure and looping over links in order of their weight. If the link connects two vertices belonging to different sets, the link is selected to be part of the minimum spanning tree, and the two sets are joined, else the link is ignored. The Kruskal algorithm assumes an undirected graph, for example, all links are bidirectional, and the weight of a link is the same in both directions:

Listing 3.8: in file: `nlib.py`

```
1  def Kruskal(graph):
2      vertices, links = graph
3      A = []
4      S = DisjointSets(len(vertices))
5      links.sort(cmp=lambda a,b: cmp(a[2],b[2]))
6      for source,dest,length in links:
7          if S.join(source,dest):
8              A.append((source,dest,length))
9      return A
```

The Kruskal algorithm goes as follows:

$$T_{worst} \in \Theta(n_E \log n_V) \qquad (3.82)$$

$$T_{memory} \in \Theta(n_E) \qquad (3.83)$$

We provide an example of application in the next subsection.

3.7.5 Minimum spanning tree: Prim

The Prim [19] algorithm solves the same problem as the Kruskal algorithm, but the Prim algorithm works on a directed graph. It works by placing all vertices in a minimum priority queue where the queue metric for each vertex is the length, or weighted value, of a link connecting the vertex to the closest known neighbor vertex. At each iteration, the algorithm pops a vertex from the priority queue, loops over its neighbors (adjacent links), and, if it finds that one of its neighbors is already in the queue and it is possible to connect it to the current vertex using a shorter link than the one connecting the neighbor to its current closest vertex, the neighbor information is then updated. The algorithm loops until there are no vertices in the priority queue.

The Prim algorithm also differs from the Kruskal algorithm because the former needs a starting vertex, whereas the latter does not. The result when interpreted as a subgraph does not depend on the starting vertex:

Listing 3.9: in file: nlib.py

```
class PrimVertex(object):
    INFINITY = 1e100
    def __init__(self,id,links):
        self.id = id
        self.closest = None
        self.closest_dist = PrimVertex.INFINITY
        self.neighbors = [link[1:] for link in links if link[0]==id]
    def __cmp__(self,other):
        return cmp(self.closest_dist, other.closest_dist)

def Prim(graph, start):
    from heapq import heappush, heappop, heapify
    vertices, links = graph
    P = [PrimVertex(i,links) for i in vertices]
    Q = [P[i] for i in vertices if not i==start]
    vertex = P[start]
    while Q:
        for neighbor_id,length in vertex.neighbors:
            neighbor = P[neighbor_id]
            if neighbor in Q and length<neighbor.closest_dist:
                neighbor.closest = vertex
                neighbor.closest_dist = length
        heapify(Q)
        vertex = heappop(Q)
    return [(v.id,v.closest.id,v.closest_dist) for v in P if not v.id==start]
```

```
>>> vertices = xrange(10)
>>> links = [(i,j,abs(math.sin(i+j+1))) for i in vertices for j in vertices]
>>> graph = [vertices,links]
>>> link = Prim(graph,0)
>>> for link in links: print link
(1, 4, 0.279...)
(2, 0, 0.141...)
(3, 2, 0.279...)
(4, 1, 0.279...)
(5, 0, 0.279...)
(6, 2, 0.412...)
(7, 8, 0.287...)
(8, 7, 0.287...)
(9, 6, 0.287...)
```

The Prim algorithm, when using a priority queue for Q, goes as follows:

$$T_{worst} \in \Theta(n_E + n_V \log n_V) \tag{3.84}$$

$$T_{memory} \in \Theta(n_E) \tag{3.85}$$

One important application of the minimum spanning tree is in evolutionary biology. Consider, for example, the DNA for the genes that produce hemoglobin, a molecule responsible for the transport of oxygen in blood. This protein is present in every animal, and the gene is also present in the DNA of every known animal. Yet its DNA structure is a little different. One can select a pool of animals and, for each two of them, compute the similarity of the DNA of their hemoglobin genes using the lcs algorithm discussed later. One can then link each two animals by a metric that represents how similar the two animals are. We can then run the Prim or the Kruskal algorithm to find the minimum spanning tree. The tree represents the most likely evolutionary tree connecting those animal species. Actually, three genes are responsible for hemoglobin (*HBA1*, *HBA2*, and *HBB*). By performing the analysis on different genes and comparing the results, it is possible to establish a consistency check of the results. [20]

Similar studies are performed routinely in evolutionary biology. They can also be applied to viruses to understand how viruses evolved over time. [21]

3.7.6   Single-source shortest paths: Dijkstra

The Dijkstra [22] algorithm solves a similar problem to the Kruskal and Prim algorithms. Given a graph, it computes, for each vertex, the shortest path connecting the vertex to a starting (or source, or root) vertex. The collection of links on all the paths defines the *single-source shortest paths*.

It works, like Prim, by placing all vertices in a min priority queue where the queue metric for each vertex is the length of the path connecting the vertex to the source. At each iteration, the algorithm pops a vertex from the priority queue, loops over its neighbors (adjacent links), and, if it finds that one of its neighbors is already in the queue and it is possible to connect it to the current vertex using a link that makes the path to the source shorter, the neighbor information is updated. The algorithm loops until there are no more vertices in the priority queue.

The implementation of this algorithm is almost identical to the Prim algorithm, except for two lines:

Listing 3.10: in file: nlib.py

```python
def Dijkstra(graph, start):
    from heapq import heappush, heappop, heapify
    vertices, links = graph
    P = [PrimVertex(i,links) for i in vertices]
    Q = [P[i] for i in vertices if not i==start]
    vertex = P[start]
    vertex.closest_dist = 0
    while Q:
        for neighbor_id,length in vertex.neighbors:
            neighbor = P[neighbor_id]
            dist = length+vertex.closest_dist
            if neighbor in Q and dist<neighbor.closest_dist:
                neighbor.closest = vertex
                neighbor.closest_dist = dist
        heapify(Q)
        vertex = heappop(Q)
    return [(v.id,v.closest.id,v.closest_dist) for v in P if not v.id==start]
```

Listing 3.11: in file: nlib.py

```python
>>> vertices = xrange(10)
>>> links = [(i,j,abs(math.sin(i+j+1))) for i in vertices for j in vertices]
>>> graph = [vertices,links]
>>> links = Dijkstra(graph,0)
>>> for link in links: print link
(1, 2, 0.897...)
(2, 0, 0.141...)
(3, 2, 0.420...)
(4, 2, 0.798...)
(5, 0, 0.279...)
(6, 2, 0.553...)
(7, 2, 0.685...)
(8, 0, 0.412...)
(9, 0, 0.544...)
```

The Dijkstra algorithm goes as follows:

$$T_{worst} \in \Theta(n_E + n_V \log n_V) \tag{3.86}$$

$$T_{memory} \in \Theta(n_E) \tag{3.87}$$

An application of the Dijkstra is in solving a maze such as the one built when discussing disjoint sets. To use the Dijkstra algorithm, we need to generate a maze, take the links representing torn-down walls, and use them to build an undirected graph. This is done by symmetrizing the links (if i and j are connected, j and i are also connected) and adding to each link a length (1, because all links connect next-neighbor cells):

```
1 >>> n,d = 4, 2
2 >>> walls, links = make_maze(n,d)
3 >>> symmetrized_links = [(i,j,1) for (i,j) in links]+[(j,i,1) for (i,j) in links
    ]
4 >>> graph = [xrange(n*n),symmetrized_links]
5 >>> links = Dijkstra(graph,0)
6 >>> paths = dict((i,(j,d)) for (i,j,d) in links)
```

Given a maze cell i, path[i] gives us a tuple (j,d) where d is the number of steps for the shortest path to reach the origin (o) and j is the ID of the next cell along this path. The following figure shows a generated maze and a reconstructed path connecting an arbitrary cell to the origin:

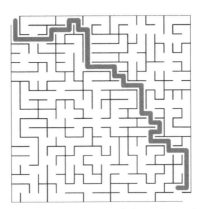

Figure 3.4: The result shows an application of the Dijkstra algorithm for the single source shortest path applied to solve a maze.

3.8   Greedy algorithms

3.8.1   Huffman encoding

The *Shannon–Fano encoding* [23][24] (also known as **minimal prefix code**) is a lossless data compression algorithm. In this encoding, each character in a string is mapped into a sequence of bits so characters that appear with less frequency are encoded with a longer sequence of bits, whereas characters that appear with more frequency are encoded with a shorter sequence.

The *Huffman encoding* [25] is an implementation of the Shannon–Fano encoding, but the sequence of bits into which each character is mapped is chosen such that the length of the compressed string is minimal. This choice is constructed in the following way. We associate a tree with each character in the string to compress. Each tree is a trivial tree containing only one node: the root node. We then associate with the root node the frequency of the character representing the tree. We then extract from the list of trees the two trees with rarest or lowest frequency: t1 and t2. We form a new tree, t3, we attach t1 and t2 to t3, and we associate a frequency with t3 equal to the sum of the frequencies of t1 and t2. We repeat this operation until the list of trees contains only one tree. At this point, we associate a sequence of bits with each node of the tree. Each bit corresponds to one level on the tree. The more frequent characters end up being closer to the root and are encoded with a few bits, while rare characters are far from the root and encoded with more bits.

PKZIP, ARJ, ARC, JPEG, MPEG3 (mp3), MPEG4, and other compressed file formats all use the Huffman coding algorithm for compressing strings. Note that Huffman is a compression algorithm with no information loss. In the JPEG and MPEG compression algorithms, Huffman algorithms are combined with some form or cut of the Fourier spectrum (e.g., MP3 is an audio compression format in which frequencies below 2 KHz are dumped and not compressed because they are not audible). Therefore the JPEG and MPEG formats are referred to as compression with information loss.

Here is a possible implementation of Huffman encoding:

Listing 3.12: in file: `nlib.py`

```
def encode_huffman(input):
    from heapq import heappush, heappop

    def inorder_tree_walk(t, key, keys):
        (f,ab) = t
        if isinstance(ab,tuple):
            inorder_tree_walk(ab[0],key+'0',keys)
            inorder_tree_walk(ab[1],key+'1',keys)
        else:
            keys[ab] = key

    symbols = {}
    for symbol in input:
        symbols[symbol] = symbols.get(symbol,0)+1
    heap = []
    for (k,f) in symbols.items():
        heappush(heap,(f,k))
    while len(heap)>1:
        (f1,k1) = heappop(heap)
        (f2,k2) = heappop(heap)
        heappush(heap,(f1+f2,((f1,k1),(f2,k2))))
    symbol_map = {}
    inorder_tree_walk(heap[0],'',symbol_map)
    encoded = ''.join(symbol_map[symbol] for symbol in input)
    return symbol_map, encoded

def decode_huffman(keys, encoded):
    reversed_map = dict((v,k) for (k,v) in keys.items())
    i, output = 0, []
    for j in xrange(1,len(encoded)+1):
        if encoded[i:j] in reversed_map:
            output.append(reversed_map[encoded[i:j]])
            i=j
    return ''.join(output)
```

We can use it as follows:

Listing 3.13: in file: `nlib.py`

```
>>> input = 'this is a nice day'
>>> keys, encoded = encode_huffman(input)
>>> print encoded
1011100111001000110010001111001010110011010000001111111110
>>> decoded = decode_huffman(keys,encoded)
>>> print decoded == input
True
```

```
8 >>> print 1.0*len(input)/(len(encoded)/8)
9 2.57...
```

We managed to compress the original data by a factor 2.57.

We can ask how good is this compression factor. The maximum theoretical best compression factor is given by the Shannon *entropy*, defined as

$$E = -\sum_u w_i \log_2 w_i \qquad (3.88)$$

where w_i is the relative frequency of each symbol. In our case, this is easy to compute as

Listing 3.14: in file: `nlib.py`

```
1 >>> from math import log
2 >>> input = 'this is a nice day'
3 >>> w = [1.0*input.count(c)/len(input) for c in set(input)]
4 >>> E = -sum(wi*log(wi,2) for wi in w)
5 >>> print E
6 3.23...
```

How could we have done better? Notice for example that the Huffman encoding does not take into account the order in which symbols appear. The original string contains the triple "is" twice, and we could have taken advantage of that pattern, but we did not.

Our choice of using characters as symbols is arbitrary. We could have used a couple of characters as symbols or triplets or any other subsequences of bytes of the original input. We could also have used symbols of different lengths for different parts of the input (we could have used a single symbol for "is"). A different choice would have given a different compression ratio, perhaps better, perhaps worse.

3.8.2 Longest common subsequence

Given two sequences of characters S_1 and S_2, this is the problem of determining the length of the longest common subsequence (LCS) that is a subsequence of both S_1 and S_2.

There are several applications for the LCS [26] algorithm:

- **Molecular biology**: DNA sequences (genes) can be represented as sequences of four letters ACGT, corresponding to the four sub-molecules forming DNA. When biologists find a new sequence, they want to find similar sequences or ones that are close. One way of computing how similar two sequences are is to find the length of their LCS.

- **File comparison**: The Unix program diff is used to compare two different versions of the same file, to determine what changes have been made to the file. It works by finding a LCS of the lines of the two files and displays the set of lines that have changed. In this instance of the problem, we should think of each line of a file as being a single complicated character.

- **Spelling correction**: If some text contains a word, w, that is not in the dictionary, a "close" word (e.g., one with a small edit distance to w) may be suggested as a correction. Transposition errors are common in written text. A transposition can be treated as a deletion plus an insertion, but a simple variation on the algorithm can treat a transposition as a single point mutation.

- **Speech recognition**: Algorithms similar to the LCS are used in some speech recognition systems—find a close match between a new utterance and one in a library of classified utterances.

Let's start with some simple observations about the LCS problem. If we have two strings, say, "ATGGCACTACGAT" and "ATCGAGC," we can represent a subsequence as a way of writing the two so that certain letters line up:

```
1    ATGGCACTACGAT
2    || |  |    |
3    ATCG AG   C
```

From this we can observe the following simple fact: if the two strings start with the same letter, it's always safe to choose that starting letter as the first character of the subsequence. This is because, if you have some other subsequence, represented as a collection of lines as drawn here, you can "push" the leftmost line to the start of the two strings without causing any other crossings and get a representation of an equally long subsequence that does start this way.

Conversely, suppose that, like in the preceding example, the two first characters differ. Then it is not possible for both of them to be part of a common subsequence. There are three possible choices: remove the first letter from either one of the strings or remove the letter from both strings.

Finally, observe that once we've decided what to do with the first characters of the strings, the remaining subproblem is again a LCS problem on two shorter strings. Therefore we can solve it recursively. However, because we don't know which choice of the three to take, we will take them all and see which choice returns the best result.

Rather than finding the subsequence itself, it turns out to be more efficient to find the length of the longest subsequence. Then, in the case where the first characters differ, we can determine which subproblem gives the correct solution by solving both and taking the max of the resulting subsequence lengths. Once we turn this into a dynamic programming algorithm, we get the following:

Listing 3.15: in file: `nlib.py`

```
def lcs(a, b):
    previous = [0]*len(a)
    for i,r in enumerate(a):
        current = []
        for j,c in enumerate(b):
            if r==c:
                e = previous[j-1]+1 if i*j>0 else 1
            else:
                e = max(previous[j] if i>0 else 0,
                        current[-1] if j>0 else 0)
            current.append(e)
        previous=current
    return current[-1]
```

Here is an example:

Listing 3.16: in file: `nlib.py`

```
>>> dna1 = 'ATGCTTTAGAGGATGCGTAGATAGCTAAATAGCTCGCTAGA'
>>> dna2 = 'GATAGGTACCACAATAATAAGGATAGCTCGCAAATCCTCGA'
>>> print lcs(dna1,dna2)
26
```

The algorithms can be shown to be $O(nm)$ (where $m = $ len(a) and $n = $ len(b)).

Another application of this algorithm is in the Unix diff utility. Here is a simple example to find the number of common lines between two files:

```
>>> a = open('file1.txt').readlines()
>>> b = open('file2.txt').readlines()
>>> print lcs(a,b)
```

3.8.3 Needleman–Wunsch

With some minor changes to the LCS algorithm, we obtain the Needleman–Wunsch algorithm [27], which solves the problem of *global sequence alignment*. The changes are that, instead of using only two alternating rows (c and d for storing the temporary results, we store all temporary results in an array z; when two matching symbols are found and they are not consecutive, we apply a penalty equal to p^m, where m is the distance between the two matches and is also the size of the gap in the matching subsequence:

Listing 3.17: in file: nlib.py

```
def needleman_wunsch(a,b,p=0.97):
    z=[]
    for i,r in enumerate(a):
        z.append([])
        for j,c in enumerate(b):
            if r==c:
                e = z[i-1][j-1]+1 if i*j>0 else 1
            else:
                e = p*max(z[i-1][j] if i>0 else 0,
                          z[i][j-1] if j>0 else 0)
            z[-1].append(e)
    return z
```

This algorithm can be used to identify common subsequences of DNA between chromosomes (or in general common similar subsequences between any two strings of binary data). Here is an example in which we look for common genes in two randomly generated chromosomes:

Listing 3.18: in file: nlib.py

```
>>> bases = 'ATGC'
```

```
2  >>> from random import choice
3  >>> genes = [''.join(choice(bases) for k in xrange(10)) for i in xrange(20)]
4  >>> chromosome1 = ''.join(choice(genes) for i in xrange(10))
5  >>> chromosome2 = ''.join(choice(genes) for i in xrange(10))
6  >>> z = needleman_wunsch(chromosome1, chromosome2)
7  >>> Canvas(title='Needleman-Wunsch').imshow(z).save('images/needleman.png')
```

The output of the algorithm is the following image:

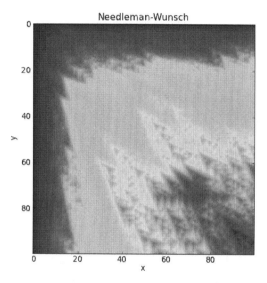

Figure 3.5: A Needleman and Wunsch plot sequence alignment. The arrow-like patterns indicate the point in the two sequences (represented by the X- and Y-coordinates) where the two sequences are more likely to align.

The arrow-like patterns in the figure correspond to locations where chromosome1 (Y coordinate) and where chromosome2 (X coordinate) have DNA in common. Those are the places where the sequences are more likely to be aligned for a more detailed comparison.

3.8.4   Continuous Knapsack

Assume you want to fill your knapsack such that you will maximize the value of its contents [28]. However, you are limited by the volume your knapsack can hold. In the continuous knapsack, the amount of each product can vary continuously. In the discrete one, each product has a finite size, and you either carry it or no.

The continuous knapsack problem can be formulated as the problem of maximizing

$$f(x) = a_0 x_0 + a_1 x_1 + \ldots + a_n x_n \tag{3.89}$$

given the constraint

$$b_0 x_0 + b_1 x_1 + \ldots + b_n x_n \leq c \tag{3.90}$$

where coefficients a_i, b_i, and c are provided and $x_i \in [0, 1]$ are to be determined.

Using financial terms, we can say that

- The set $\{x_0, x_1, \ldots, x_n\}$ forms a portfolio

- b_i is the cost of investment i

- c is the total investment capital available

- a_i is the expected return of investment for investment i

- $f(x)$ is the expected value of our portfolio $\{x_0, x_1, \ldots, x_n\}$

Here is the solving algorithm:

Listing 3.19: in file: `nlib.py`

```
def continuum_knapsack(a,b,c):
    table = [(a[i]/b[i],i) for i in xrange(len(a))]
    table.sort()
    table.reverse()
    f=0.0
    for (y,i) in table:
        quantity = min(c/b[i],1)
        x.append((i,quantity))
        c = c-b[i]*quantity
        f = f+a[i]*quantity
    return (f,x)
```

This algorithm is dominated by the sort; therefore

$$T_{worst}(x) \in O(n \log n) \tag{3.91}$$

3.8.5   Discrete Knapsack

The discrete Knapsack problem is very similar to the continuous knapsack problem but $x_i \in \{0, 1\}$ (can only be 0 or 1).

Consider the jars of liquids replaced with baskets of objects, say, a basket each of gold bars, silver coins, copper beads, and Rolex watches. How many of each item do you take? The discrete knapsack problem does not consider "baskets of items" but rather all the items together. In this example, dump out all the baskets and you have individual objects to take. Which objects do you take, and which do you leave behind?

In this case, a greedy approach does not apply and the problem is, in general, NP complete. This concept is defined formally later but it means that there is no known algorithm that can solve this problem and that its order of growth is a polynomial. The best known algorithm has an exponential running time.

This kind of problem is unsolvable for large input.

If we assume that c and b_i are all multiples of a finite factor ε, then it is possible to solve the problem in $O(c/\varepsilon)$. Even when there is not a finite factor ε, we can always round c and b_i to some finite precision ε, and we can conclude that, for any finite precision ε, we can solve the problem in linear time. The algorithm that solves this problem follows a dynamic programming approach.

We can reformulate the problem in terms of a simple capital budgeting problem. We have to invest \$5M. We assume $\varepsilon =$\$1M. We are in contact with three investment firms. Each offers a number of investment opportunities characterized by an investment cost $c[i, j]$ and an expected return of investment $r[i, j]$. The index i labels the investment firm and the index j labels the different investment opportunities offered by the firm. We have to build a portfolio that maximizes the return of investment. We cannot select more than one investment for each firm, and we cannot select fractions of investments.

Without loss of generality, we will assume that

$$c[i,j] \leq c[i,j+1] \text{ and } r[i,j] \leq r[i,j+1] \qquad (3.92)$$

which means that investment opportunities for each firm are sorted according to their cost.

Consider the following explicit case:

	Firm	$i = 0$	Firm	$i = 1$	Firm	$i = 2$	
proposal	$c[0,j]$	$r[0,j]$	$c[1,j]$	$r[1,j]$	$c[2,j]$	$r[2,j]$	
$j = 0$	0	0	0	0	0	0	(Table 1)
$j = 1$	1	5	2	8	1	4	
$j = 2$	2	6	3	9	-	-	
$j = 3$	-	-	4	12	-	-	

(table values are always multiples of $\varepsilon = \$1M$).

Notice that we can label each possible portfolio by a triplet $\{j_0, j_1, j_2\}$.

A straightforward way to solve this is to try all possibilities and choose the best. In this case, there are only $3 \times 4 \times 2 = 24$ possible portfolios. Many of these are infeasible (e.g., portfolio $\{2, 3, 0\}$ costs $\$6M$ and we cannot afford it). Other portfolios are feasible but very poor (like portfolio $\{0, 0, 1\}$, which is feasible but returns only $\$4M$).

Here are some disadvantages of total enumeration:

- For larger problems, the enumeration of all possible solutions may not be computationally feasible.

- Infeasible combinations may not be detectable a priori, leading to inefficiency.

- Information about previously investigated combinations is not used to eliminate inferior or infeasible combinations (unless we use memoization, but in this case the algorithm would grow polynomially in memory space).

We can, instead, use a dynamic programming approach.

We break the problem into three stages, and at each stage, we fill a table of optimal investments for each discrete amount of money. At each stage i, we only consider investments from firm i and the table during the previous stage.

So stage 0 represents the money allocated to firm 0, stage 1 the money to firm 1, and stage 2 the money to firm 2.

STAGE ZERO: we maximize the return of investment considering only offers from firm o. We fill a table $f[0,k]$ with the maximum return of investment if we invest k million dollars in firm 0:

$$f[0,k] = \max_{j|c[0,j]<k} r[0,j] \tag{3.93}$$

k	$f[0,k]$
0	0
1	5
2*	6*
3	6
4	6
5	6

(3.94)

STAGE TWO: we maximize the return of investment considering offers from firm 1 and the prior table. We fill a table $f[1,k]$ with the maximum return of investment if we invest k million dollars in firm 0 and firm 1:

$$f[1,k] = \max_{j|c[1,j]<k} r[1,j] + f[0,k-c[0,j]] \tag{3.95}$$

k	$c[2,j]$	$f[0,k-c[0,j]]$	$f[1,k]$
0	0	0	0
1	0	1	5
2	2	0	8
3	2	1	9
4	3	1	13
5*	4*	1*	18*

(3.96)

STAGE THREE: we maximize the return of investment considering offers from firm 2 and the preceding table. We fill a table $f[2,k]$ with the maximum return of investment if we invest k million dollars in firm 0, firm 1, and firm 2:

$$f[2,k] = \max_{j|c[2,j]<k} r[2,j] + f[1,k-c[1,j]] \tag{3.97}$$

k	$c[2,j]$	$f[1,k-c[1,j]]$	$f[2,k]$
0	0	0	0
1	0	1	5
2	2	0	8
3	2	1	9
4	1	3	13
5^*	2^*	3^*	18^*

$$(3.98)$$

The maximum return of investment with \$5M is therefore \$18M. It can be achieved by investing \$2M in firm 2 and \$3M in firms 0 and 1. The optimal choice is marked with a star in each table. Note that to determine how much money has to be allocated to maximize the return of investment requires storing past tables to be able to look up the solution to subproblems.

We can generalize eq.(3.95) and eq.(3.97) for any number of investment firms (decision stages):

$$f[i,k] = \max_{j|c[i,j]<k} r[i,j] + f[i-1,k-c[i-1,j]] \tag{3.99}$$

3.9 Artificial intelligence and machine learning

3.9.1 Clustering algorithms

There are many algorithms available to cluster data [29]. They are all based on empirical principles because the cluster themselves are defined by the algorithm used to identify them. Normally we distinguish three categories:

- *Hierarchical clustering*: These algorithms start by considering each point a cluster of its own. At each iteration, the two clusters closest to each other are joined together, forming a larger cluster. Hierarchical clustering algorithms differ from each other about the rule used to determine the distance between clusters. The algorithm returns a tree representing the clusters that are joined, called a *dendrogram*.

- *Centroid-based clustering*: These algorithms require that each point be represented by a vector and each cluster also be represented by a vector (centroid of the cluster). With each iteration, a better estimation for the centroids is given. An example of centroid-based clustering is *k-means* clustering. These algorithms require an a priori knowledge of the number of clusters and return the position of the centroids as well the set of points belonging to each cluster.

- *Distribution-based clustering*: These algorithms are based on statistics (more than the other two categories). They assume the points are generated from a distribution (which mush be known a priori) and determine the parameters of the distribution. It provides clustering because the distribution may be a sum of more than one localized distribution (each being a cluster).

Both k-means and distribution-based clustering assume an a priori knowledge about the data that often defies the purpose of using clustering: learn something we do now know about the data using an empirical algorithm. They also require that the points be represented by vectors in a Euclidean space, which is not always the case. Consider the case of clustering DNA sequences or financial time series. Technically the latter can be presented as vectors, but their dimensionality can be very large, thus making the algorithms impractical.

Hierarchical clustering only requires the notion of a distance between points, for some of the points.

The following algorithm is a hierarchical clustering algorithm with the following characteristics:

- Individual points do not need to be vectors (although they can be).

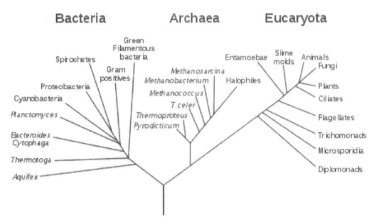

Figure 3.6: Example of a dendrogram.

- Points may have a weight used to determine their relative importance in identifying the characteristics of the cluster (think of clustering financial assets based on the time series of their returns; the weight could the average traded volume).

- The distance between points is computed by a metric function provided by the user. The metric can return None if there is no known connection between two points.

- The algorithm can be used to build the entire *dendrogram* , or it can stop for a given value of k, a target number of clusters.

- For points that are vectors and a given k, the result is similar to the result of the k-means clustering.

The algorithm works like any other hierarchical clustering algorithm. At the beginning, all-to-all distances are computed and stored in a list d. Each point is its own cluster. At each iteration, the two clusters closer together are merged to form one bigger cluster. The distance between each other cluster and the merged cluster is computed by performing a weighted average of the distances between the other cluster and the two merged clusters. The weight factors are provided as input. This is equivalent to what the k-means algorithm does by computing the position of a centroid based on the vectors of the member points.

The algorithm self.q implements disjointed sets representing the set of clusters. The algorithm self.q is a dictionary. If self.q[i] is a list, then i is its own cluster, and the list contains the IDs of the member points. If self.q[i] is an integer, then cluster i is no longer its own cluster as it was merged to the cluster represented by the integer.

At each point in time, each cluster is represented by one element, which can be found recursively by self.parent(i). This function returns the ID of the cluster containing element i and returns a list of IDs of all points in the same cluster:

Listing 3.20: in file: nlib.py

```
class Cluster(object):
    def __init__(self,points,metric,weights=None):
        self.points, self.metric = points, metric
        self.k = len(points)
        self.w = weights or [1.0]*self.k
        self.q = dict((i,[i]) for i,e in enumerate(points))
        self.d = []
        for i in xrange(self.k):
            for j in xrange(i+1,self.k):
                m = metric(points[i],points[j])
                if not m is None:
                    self.d.append((m,i,j))
        self.d.sort()
        self.dd = []
    def parent(self,i):
        while isinstance(i,int): (parent, i) = (i, self.q[i])
        return parent, i
    def step(self):
        if self.k>1:
            # find new clusters to join
            (self.r,i,j),self.d = self.d[0],self.d[1:]
            # join them
            i,x = self.parent(i) # find members of cluster i
            j,y = self.parent(j) # find members if cluster j
            x += y               # join members
            self.q[j] = i        # make j cluster point to i
            self.k -= 1          # decrease cluster count
            # update all distances to new joined cluster
            new_d = [] # links not related to joined clusters
```

```
30        old_d = {} # old links related to joined clusters
31        for (r,h,k) in self.d:
32            if h in (i,j):
33                a,b = old_d.get(k,(0.0,0.0))
34                old_d[k] = a+self.w[k]*r,b+self.w[k]
35            elif k in (i,j):
36                a,b = old_d.get(h,(0.0,0.0))
37                old_d[h] = a+self.w[h]*r,b+self.w[h]
38            else:
39                new_d.append((r,h,k))
40        new_d += [(a/b,i,k) for k,(a,b) in old_d.items()]
41        new_d.sort()
42        self.d = new_d
43        # update weight of new cluster
44        self.w[i] = self.w[i]+self.w[j]
45        # get new list of cluster members
46        self.v = [s for s in self.q.values() if isinstance(s,list)]
47        self.dd.append((self.r,len(self.v)))
48        return self.r, self.v
49
50    def find(self,k):
51        # if necessary start again
52        if self.k<k: self.__init__(self.points,self.metric)
53        # step until we get k clusters
54        while self.k>k: self.step()
55        # return list of cluster members
56        return self.r, self.v
```

Given a set of points, we can determine the most likely number of clusters representing the data, and we can make a plot of the number of clusters versus distance and look for a plateau in the plot. In correspondence with the plateau, we can read from the y-coordinate the number of clusters. This is done by the function cluster in the preceding algorithm, which returns the average distance between clusters and a list of clusters.

For example:

Listing 3.21: in file: nlib.py

```
1 >>> def metric(a,b):
2 ...     return math.sqrt(sum((x-b[i])**2 for i,x in enumerate(a)))
3 >>> points = [[random.gauss(i % 5,0.3) for j in xrange(10)] for i in xrange(200)]
4 >>> c = Cluster(points,metric)
5 >>> r, clusters = c.find(1) # cluster all points until one cluster only
6 >>> Canvas(title='clustering example',xlab='distance',ylab='number of clusters'
7 ...     ).plot(c.dd[150:]).save('clustering1.png')
```

```
>>> Canvas(title='clustering example (2d projection)',xlab='p[0]',ylab='p[1]'
...         ).ellipses([p[:2] for p in points]).save('clustering2.png')
```

With our sample data, we obtain the following plot ("clustering1.png"):

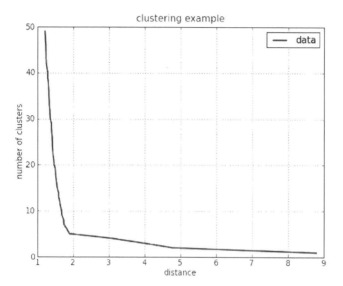

Figure 3.7: Number of clusters found as a function of the distance cutoff.

and the location where the curve bends corresponds to five clusters. Although our points live in 10 dimensions, we can try to project them into two dimensions and see the five clusters ("clustering2.png"):

3.9.2   Neural network

An artificial *neural network* is an electrical circuit (usually simulated in software) that mimics the functionality of the neurons in the animal (and human) brain [30]. It is usually employed in pattern recognition. The network consists of a set of simulated neurons, connected by links (synapses). Some links connect the neurons with each other, some connect the neurons with the input and some with the output. Neurons are usually organized in the layers with one *input layer* of neurons connected only with the input and the next layer. Another one, the *output layer*, comprises neu-

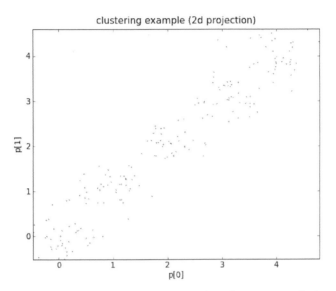

Figure 3.8: Visual representation of the clusters where the points coordinates are projected in 2D.

rons connected only with the output and previous layers, or many *hidden layers* of neurons connected only with other neurons. Each neuron is characterized by input links and output links. Each output of a neuron is a function of its inputs. The exact shape of that function depends on the network and on parameters that can be adjusted. Usually this function is chosen to be a monotonic increasing function on the sum of the inputs, where both the inputs and the outputs take values in the [0,1] range. The inputs can be thought as electrical signals reaching the neuron. The output is the electrical signal emitted by the neuron. Each neuron is defined by a set of parameters a which determined the relative weight of the input signals. A common choice for this characteristic function is:

$$\text{output}_{ij} = \tanh(\sum_k a_{ijk}\text{input}_{ik}) \qquad (3.100)$$

where i labels the neuron, j labels the output, k labels the input, and a_{ijk} are characteristic parameters describing the neurons.

The network is trained by providing an input and adjusting the characteristics a_{ijk} of each neuron k to produce the expected output. The network is trained iteratively until its parameters converge (if they converge), and then it is ready to make predictions. We say the network has learned from the training data set.

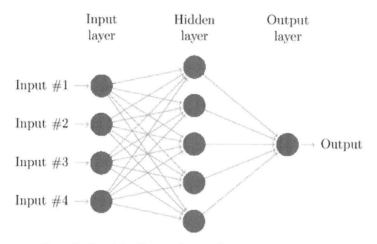

Figure 3.9: Example of a minimalist neural network.

Listing 3.22: in file: nlib.py

```python
class NeuralNetwork:
    """
    Back-Propagation Neural Networks
    Placed in the public domain.
    Original author: Neil Schemenauer <nas@arctrix.com>
    Modified by: Massimo Di Pierro
    Read more: http://www.ibm.com/developerworks/library/l-neural/
    """

    @staticmethod
    def rand(a, b):
        """ calculate a random number where:  a <= rand < b """
        return (b-a)*random.random() + a

    @staticmethod
    def sigmoid(x):
        """ our sigmoid function, tanh is a little nicer than the standard 1/(1+
            e^-x) """
        return math.tanh(x)
```

```
@staticmethod
def dsigmoid(y):
    """ # derivative of our sigmoid function, in terms of the output """
    return 1.0 - y**2

def __init__(self, ni, nh, no):
    # number of input, hidden, and output nodes
    self.ni = ni + 1 # +1 for bias node
    self.nh = nh
    self.no = no

    # activations for nodes
    self.ai = [1.0]*self.ni
    self.ah = [1.0]*self.nh
    self.ao = [1.0]*self.no

    # create weights
    self.wi = Matrix(self.ni, self.nh, fill=lambda r,c: self.rand(-0.2, 0.2)
        )
    self.wo = Matrix(self.nh, self.no, fill=lambda r,c: self.rand(-2.0, 2.0)
        )

    # last change in weights for momentum
    self.ci = Matrix(self.ni, self.nh)
    self.co = Matrix(self.nh, self.no)

def update(self, inputs):
    if len(inputs) != self.ni-1:
        raise ValueError('wrong number of inputs')

    # input activations
    for i in xrange(self.ni-1):
        self.ai[i] = inputs[i]

    # hidden activations
    for j in xrange(self.nh):
        s = sum(self.ai[i] * self.wi[i,j] for i in xrange(self.ni))
        self.ah[j] = self.sigmoid(s)

    # output activations
    for k in xrange(self.no):
        s = sum(self.ah[j] * self.wo[j,k] for j in xrange(self.nh))
        self.ao[k] = self.sigmoid(s)
    return self.ao[:]

def back_propagate(self, targets, N, M):
    if len(targets) != self.no:
        raise ValueError('wrong number of target values')
```

```
66
67        # calculate error terms for output
68        output_deltas = [0.0] * self.no
69        for k in xrange(self.no):
70            error = targets[k]-self.ao[k]
71            output_deltas[k] = self.dsigmoid(self.ao[k]) * error
72
73        # calculate error terms for hidden
74        hidden_deltas = [0.0] * self.nh
75        for j in xrange(self.nh):
76            error = sum(output_deltas[k]*self.wo[j,k] for k in xrange(self.no))
77            hidden_deltas[j] = self.dsigmoid(self.ah[j]) * error
78
79        # update output weights
80        for j in xrange(self.nh):
81            for k in xrange(self.no):
82                change = output_deltas[k]*self.ah[j]
83                self.wo[j,k] = self.wo[j,k] + N*change + M*self.co[j,k]
84                self.co[j,k] = change
85                #print N*change, M*self.co[j,k]
86
87        # update input weights
88        for i in xrange(self.ni):
89            for j in xrange(self.nh):
90                change = hidden_deltas[j]*self.ai[i]
91                self.wi[i,j] = self.wi[i,j] + N*change + M*self.ci[i,j]
92                self.ci[i,j] = change
93
94        # calculate error
95        error = sum(0.5*(targets[k]-self.ao[k])**2 for k in xrange(len(targets))
              )
96        return error
97
98    def test(self, patterns):
99        for p in patterns:
100            print p[0], '->', self.update(p[0])
101
102    def weights(self):
103        print 'Input weights:'
104        for i in xrange(self.ni):
105            print self.wi[i]
106        print
107        print 'Output weights:'
108        for j in xrange(self.nh):
109            print self.wo[j]
110
111    def train(self, patterns, iterations=1000, N=0.5, M=0.1, check=False):
112        # N: learning rate
113        # M: momentum factor
```

```
114     for i in xrange(iterations):
115         error = 0.0
116         for p in patterns:
117             inputs = p[0]
118             targets = p[1]
119             self.update(inputs)
120             error = error + self.back_propagate(targets, N, M)
121         if check and i % 100 == 0:
122             print 'error %-14f' % error
```

In the following example, we teach the network the XOR function, and we create a network with two inputs, two intermediate neurons, and one output. We train it and check what it learned:

Listing 3.23: in file: nlib.py

```
1  >>> pat = [[[0,0], [0]], [[0,1], [1]], [[1,0], [1]], [[1,1], [0]]]
2  >>> n = NeuralNetwork(2, 2, 1)
3  >>> n.train(pat)
4  >>> n.test(pat)
5  [0, 0] -> [0.00...]
6  [0, 1] -> [0.98...]
7  [1, 0] -> [0.98...]
8  [1, 1] -> [-0.00...]
```

Now, we use our neural network to learn patterns in stock prices and predict the next day return. We then check what it has learned, comparing the sign of the prediction with the sign of the actual return for the same days used to train the network:

Listing 3.24: in file: test.py

```
1  >>> storage = PersistentDictionary('sp100.sqlite')
2  >>> v = [day['arithmetic_return']*300 for day in storage['AAPL/2011'][1:]]
3  >>> pat = [[v[i:i+5],[v[i+5]]] for i in xrange(len(v)-5)]
4  >>> n = NeuralNetwork(5, 5, 1)
5  >>> n.train(pat)
6  >>> predictions = [n.update(item[0]) for item in pat]
7  >>> success_rate = sum(1.0 for i,e in enumerate(predictions)
8  ...                    if e[0]*v[i+5]>0)/len(pat)
```

The learning process depends on the random number generator; therefore, sometimes, for this small training data set, the network succeeds in predicting the sign of the next day arithmetic return of the stock with more than 50% probability, and sometimes it does not. We leave it to the reader to study the significance of this result but using a different subset of the data for the training of the network and for testing its success rate.

3.9.3   Genetic algorithms

Here we consider a simple example of genetic algorithms [31].

We have a population of chromosomes in which each chromosome is just a data structure, in our example, a string of random "ATGC" characters.

We also have a metric to measure the fitness of each chromosome.

At each iteration, only the top-ranking chromosomes in the population survive. The top 10 mate with each other, and their offspring constitute the population for the next iteration. When two members of the population mate, the newborn member of the population has a new DNA sequence, half of which comes from the father and half from the mother, with two randomly mutated DNA basis.

The algorithm stops when we reach a maximum number of generations or we find a chromosome of the population with maximum fitness.

In the following example, the fitness is measured by the similarity between a chromosome and a random target chromosome. The population evolves to approximate better and better that one random target chromosome:

```
 1  from random import randint, choice
 2
 3  class Chromosome:
 4      alphabet = 'ATGC'
 5      size = 32
 6      mutations = 2
 7      def __init__(self,father=None,mother=None):
 8          if not father or not mother:
 9              self.dna = [choice(self.alphabet) for i in xrange(self.size)]
10          else:
11              self.dna = father.dna[:self.size/2]+mother.dna[self.size/2:]
12              for mutation in xrange(self.mutations):
13                  self.dna[randint(0,self.size-1)] = choice(self.alphabet)
14      def fitness(self,target):
15          return sum(1 for i,c in enumerate(self.dna) if c==target.dna[i])
16
17  def top(population,target,n=10):
18      table = [(chromo.fitness(target), chromo) for chromo in population]
19      table.sort(reverse = True)
20      return [row[1] for row in table][:n]
```

```
21
22  def oneof(population):
23      return population[randint(0, len(population)-1)]
24
25  def main():
26      GENERATIONS = 10000
27      OFFSPRING = 20
28      SEEDS = 20
29      TARGET = Chromosome()
30
31      population = [Chromosome() for i in xrange(SEEDS)]
32      for i in xrange(GENERATIONS):
33          print '\n\nGENERATION:',i
34          print 0, TARGET.dna
35          fittest = top(population,TARGET)
36          for chromosome in fittest: print i,chromosome.dna
37          if max(chromo.fitness(TARGET) for chromo in fittest)==Chromosome.size:
38              print 'SOLUTION FOUND'
39              break
40          population = [Chromosome(father=oneof(fittest),mother=oneof(fittest)) \
41                        for i in xrange(OFFSPRING)]
42
43  if __name__=='__main__': main()
```

Notice that this algorithm can easily be modified to accommodate other fitness metrics and DNA that consists of a data structure other than a sequence of "ATGC" symbols. The only trickery is finding a proper mating algorithm that preserves some of the fitness features of the parents in the DNA of their offspring. If this does not happen, each next generation loses the fitness properties gained by its parents, thus causing the algorithm not to converge. In our case, it works because if the parents are "close" to the target, then half of the DNA of each parent is also close to the corresponding half of the target DNA. Therefore the DNA of the offspring is as fit as the average of their parents. On top of this, the two random mutations allow the algorithm to further explore the space of all possible DNA sequences.

3.10 Long and infinite loops

3.10.1   P, NP, and NPC

We say a problem is in P if it can be solved in polynomial time: $T_{worst} \in O(n^\alpha)$ for some α.

We say a problem is in NP if an input string can be verified to be a solution in polynomial time: $T_{worst} \in O(n^\alpha)$ for some α.

We say a problem is in co-NP if an input string can be verified not to be a solution in polynomial time: $T_{worst} \in O(n^\alpha)$ for some α.

We say a problem is in NPH (NP Hard) if it is harder than any other problem in NP.

We say a problem is in NPC (NP Complete) if it is in NP and in NPH. Consequences:

$$\text{if } \exists x \mid x \in NPC \text{ and } x \in P \Rightarrow \forall y \in NP, y \in P \qquad (3.101)$$

There are a number of open problems about the relations among these sets. Is the set co-NP equivalent to NP? Or perhaps is the intersection between co-NP and NP equal to P? Are NP and NPC the same set? These questions are very important in computer science because if, for example, NP turns out to be the same set as NPC, it means that it must be possible to find algorithms that solve in polynomial time problems that currently do not have a polynomial time solution. Conversely, if one could prove that NP is not equivalent to NPC, we would know that a polynomial time solution to NPC problems does not exist [32].

3.10.2   Cantor's argument

Cantor proved that the real numbers in any interval (e.g., in $[0, 1)$) are more than the integer numbers, therefore real numbers are uncountable [33]. The proof proceeds as follows:

1. Consider the real numbers in the interval $[0, 1)$ not including 1.

2. Assume that these real numbers are countable. Therefore it is possible

to associate each of them to an integer

$$
\begin{array}{ccl}
1 & \longleftrightarrow & 0.xxxxxxxxxx... \\
2 & \longleftrightarrow & 0.xxxxxxxxxx... \\
3 & \longleftrightarrow & 0.xxxxxxxxxx... \\
4 & \longleftrightarrow & 0.xxxxxxxxxx... \\
5 & \longleftrightarrow & 0.xxxxxxxxxx... \\
... & ... & ...
\end{array}
\tag{3.102}
$$

(here x represent a decimal digit of a real numbers)

3. Now construct a number $\alpha = 0.yyyyyyyy....$ where the first decimal digit differs from the first decimal digit of the first real number of table 3.102, the second decimal digit differs from the second decimal digit of the second real number of table 3.102, and so on and so on for all the infinite decimal digits:

$$
\begin{array}{ccl}
1 & \longleftrightarrow & 0.\bar{x}xxxxxxxxx... \\
2 & \longleftrightarrow & 0.x\bar{x}xxxxxxxx... \\
3 & \longleftrightarrow & 0.xx\bar{x}xxxxxxx... \\
4 & \longleftrightarrow & 0.xxx\bar{x}xxxxxx... \\
5 & \longleftrightarrow & 0.xxxx\bar{x}xxxxx... \\
... & ... & ...
\end{array}
\tag{3.103}
$$

4. The new number α is a real number, and by construction, it is not in the table. In fact, it differs with each item by at least one decimal digit. Therefore the existence of α disproves the assumption that all real numbers in the interval $[0, 1)$ are listed in the table.

There is a very practical consequence of this argument. In fact, in chapter 2, we have seen the distinction between type float and class Decimal. We have seen about pitfalls of float and how Decimal can represent floating point numbers with arbitrary precision (assuming we have the memory to do so). Cantor's argument tells us there are numbers that cannot even be represented as Decimal because they would require an infinite amount of storage; π and e are examples of these numbers.

3.10.3 Gödel's theorem

Gödel used a similar diagonal argument to prove that there are as many problems (or theorems) as real numbers and as many algorithms (or proofs) as natural numbers [33]. Because there is more of the former than the latter, it follows that there are problems for which there is no corresponding solving algorithm. Another interpretation of Gödel's theorem is that, in any formal language, for example, mathematics, there are theorems that cannot be proved.

Another consequence of Gödel's theorem is the following: it is impossible to write a computer program to test if a given algorithm stops or enters into an infinite loop.

Consider the following code:

```
def next(i):
    while len(set(str(i*i))) > 2:
        i=i+2
    print i

next(81621)
```

This code check searches for a number equal or greater than 81621 which square is comprised of only two digits. Nobody knows whether such number exists, therefore nobody knows if this code stops.

Although one day this problem may be solved, there are many other problems that are still unsolved; actually, there are an infinite number of them.

4

Numerical Algorithms

4.1 Well-posed and stable problems

Numerical algorithms deal mostly with well-posed and stable problems.

A problem is well posed if

- The solution exists and is unique

- The solution has a continuous dependence on input data (a small change in the input causes a small change in the output)

Most physical problems are well posed, except at *critical points*, where any infinitesimal variation in one of the input parameters of the system can cause a large change in the output and therefore in the behavior of the system. This is called *chaos*.

Consider the case of dropping a ball on a triangular-shaped mountain. Let the input of the problem be the horizontal position where the drop occurs and the output the horizontal position of the ball after a fixed amount of time. Almost anywhere the ball is dropped, it will roll down the mountain following deterministic and classical laws of physics, thus the position is calculable and a continuous function of the input position. This is true everywhere, except when the ball is dropped on top of the peak of the mountain. In this case, a minor infinitesimal variation to the right or to the left can make the ball roll to the right or to the left, respectively. Therefore this is not a well posed problem.

A problem is said to be *stable* if the solution is not just continuous but also weakly sensitive to input data. This means that the change of the output (in percent) is smaller than the change in the input (in percent).

Numerical algorithms work best with stable problems.

We can quantify this as follows. Let x be an input and y be the output of a function:

$$y = f(x) \tag{4.1}$$

We define the condition number of f in x as

$$\mathrm{cond}(f, x) \equiv \frac{|dy/y|}{|dx/x|} = |xf'(x)/f(x)| \tag{4.2}$$

(the latter equality only holds if f is differentiable in x).

A problem with a low condition number is said to be well-conditioned, while a problem with a high condition number is said to be ill-conditioned. XXX

We say that a problem characterized by a function f is well conditioned in a domain D if the condition number is less than 1 for every input in the domain. We also say that a problem is stable if it is well conditioned.

In this book, we are mostly concerned with stable (well-conditioned) problems. If a problem is well-conditioned in for all input in a domain, it is also stable.

4.2   Approximations and error analysis

Consider a physical quantity, for example, the length of a nail. Given one nail, we can measure its length by choosing a measuring instrument. Whatever instrument we choose, we will be able to measure the length of the nail within the resolution of the instrument. For example, with a tape measure with a resolution of 1 mm, we will only be able to determine the length of the nail within 1 mm of resolution. Repeated measurements performed at different times, by different people, using different instruments may bring different results. We can choose a more precise instrument, but it would not change the fact that different measures will bring different values compatible with the resolution of the instrument. Eventually one will have to face the fact that there may not be such a thing as the length of a nail. For example, the length varies with the temperature and the details of how the measurement is performed. In fact, a nail (as everything else) is made out of atoms, which are made of protons, neutrons, and electrons, which determine an electromagnetic cloud that fluctuates in space and time and depends on the surrounding objects and interacts with the instrument of measure. The length of the nail is the result of a measure.

For each measure there is a result, but the results of multiple measurements are not identical. The results of many measurements performed with the same resolution can be summarized in a distribution of results. This distribution will have a mean $\bar{x}$ and a standard deviation δx, which we call uncertainty. From now on, unless otherwise specified, we assume that the distribution of results is Gaussian so that $\bar{x}$ can be interpreted as the mean and δx as the standard deviation.

Now let us consider a system that, given an input x, produces the output y; x and y are physical quantities that we can measure, although only with a finite resolution. We can model the system with a function f such that $y = f(x)$ and, in general, f is not known.

We have to make various approximations:

- We can replace the "true" value for the input with our best estimate, $\bar{x}$, and its associated uncertainty, δx.

- We can replace the "true" value for the output with our best estimate, $\bar{y}$, and its associated uncertainty, δy.

- Even if we know there is a "true" function f describing the system, our implementation for the function is always an approximation, $\bar{f}$. In fact, we may not have a single approximation but a series of approximations of increasing precision, f_n, which become more and more accurate (usually) as n increases. If we are lucky, up to precision errors, as n increases, our approximations will become closer and closer to f, but this will take an infinite amount of time. We have to stop at some finite n.

With the preceding definition, we can define the following types of errors:

- **Data error**: the difference between x and $\bar{x}$.

- **Computational error**: the difference between $\bar{f}(\bar{x})$ and y. Computational error includes two parts systematic error and statistical error.

- **Statistical error**: due to the fact that, often, the computation of $\bar{f}(x) = \lim_{n \to \infty} f_n(x)$ is too computationally expensive and we must approximate $\bar{f}(x)$ with $f_n(x)$. This error can be estimated and controlled.

- **Systematic error**: due to the fact that $\bar{f}(x) = \lim_{n \to \infty} f_n(x) \neq f(x)$. This is for two reasons: modeling errors (we do not know $f(x)$) and rounding errors (we do not implement $f(x)$ with arbitrary precision arithmetics).

- **Total error**: defined as the computational error + the propagated data error and in a formula:

$$\delta y = |f(\bar{x}) - f_n(\bar{x})| + |f_n'(\bar{x})|\delta x \qquad (4.3)$$

The first term is the computational error (we use f_n instead of the true f), and the second term is the propagated data error (δx, the uncertainty in x, propagates through f_n).

4.2.1 Error propagation

When a variable x has a finite Gaussian uncertainty δx, how does the uncertainty propagate through a function f? Assuming the uncertainty is small, we can always expand using a Taylor series:

$$y + \delta y = f(x + \delta x) = f(x) + f'(x)\delta x + O(\delta x^2) \qquad (4.4)$$

And because we interpret δy as the width of the distribution y, it should be positive:

$$\delta y = |f'(x)|\delta x \tag{4.5}$$

We have used this formula before for the propagated data error. For functions of two variables $z = f(x, y)$ and assuming the uncertainties in x and y are independent,

$$\delta z = \sqrt{\left|\frac{\partial f(x, y)}{\partial x}\right|^2 \delta x^2 + \left|\frac{\partial f(x, y)}{\partial y}\right|^2 \delta y^2} \tag{4.6}$$

which for simple arithmetic operations reduces to

$$
\begin{array}{ll}
z = x + y & \delta z = \sqrt{\delta x^2 + \delta y^2} \\
z = x - y & \delta z = \sqrt{\delta x^2 + \delta y^2} \\
z = x * y & \delta z = |x * y|\sqrt{(\delta x/x)^2 + (\delta y/y)^2} \\
z = x / y & \delta z = |x/y|\sqrt{(\delta x/x)^2 + (\delta y/y)^2}
\end{array}
$$

Notice that when $z = x - y$ approaches zero, the uncertainty in z is larger than the uncertainty in x and y and can overwhelm the result. Also notice that if $z = x/y$ and y is small compared to x, then the uncertainty in z can be large. Bottom line: try to avoid differences between numbers that are in proximity of each other and try to avoid dividing by small numbers.

4.2.2   buckingham

Buckingham is a Python library that implements error propagation and unit conversion. It defines a single class called Number, and a number object has value, an uncertainty, and a dimensionality (e.g., length, volume, mass).

Here is an example:

```
>>> from buckingham import *
>>> globals().update(allunits())
>>> L = (4 + pm(0.5)) * meter
>>> v = 5 * meter/second
>>> t = L/v
>>> print t)
```

```
 7  (8.00 +/- 1.00)/10
 8  >>> print t.units()
 9  second
10  >>> print t.convert('hour')
11  (2.222 +/- 0.278)/10^4
```

Notice how adding an uncertainty to a numeric value with +
pm(...) or adding units to a numeric value (integer or floating point)
transforms the float number into a Number object. A Number object be-
haves like a floating point but propagates its uncertainty and its units.
Internally, all units are converted to the International System, unless an
explicit conversion is specified.

4.3 Standard strategies

Here are some strategies that are normally employed in numerical algorithms:

- Approximate a continuous system with a discrete system

- Replace integrals with sums

- Replace derivatives with finite differences

- Replace nonlinear with linear + corrections

- Transform a problem into a different one

- Approach the true result by iterations

Here are some examples of each of the strategies.

4.3.1 Approximate continuous with discrete

Consider a ball in a one-dimensional box of size L, and let x be the position of the ball in the box. Instead of treating x as a continuous variable,
we can assume a finite resolution of $h = L/n$ (where h is the minimum
distance we can distinguish without instruments and n is the maximum
number of distinct discrete points we can discriminate), and set $x \equiv hi$,
where i is an integer in between 0 and n; $x = 0$ when $i = 0$ and $x = L$

when $i = n$.

4.3.2 Replace derivatives with finite differences

Computing $df(x)/dx$ analytically is only possible when the function f is expressed in simple analytical terms. Computing it analytically is not possible when $f(x)$ is itself implemented as a numerical algorithm. Here is an example:

```
1  def f(x):
2      (s,t) = (1.0,1.0)
3      for i in xrange(1,10): (s, t) = (s+t, t*x/i)
4      return s
```

What is the derivative of $f(x)$?

The most common ways to define a derivative are the right derivative

$$\frac{df^+(x)}{dx} = \lim_{h \to 0} \frac{f(x+h) - f(x)}{h} \qquad (4.7)$$

the left derivative

$$\frac{df^-(x)}{dx} = \lim_{h \to 0} \frac{f(x) - f(x-h)}{h} \qquad (4.8)$$

and the average of the two

$$\frac{df(x)}{dx} = \frac{1}{2}\left(\frac{df^+(x)}{dx} + \frac{df^-(x)}{dx}\right) = \lim_{h \to 0} \frac{f(x+h) - f(x-h)}{2h} \qquad (4.9)$$

If the function is differentiable in x, then, by definition of "differentiable," the left and right definitions are equal, and the three prior definitions are equivalent. We can pick one or the other, and the difference will be a systematic error.

If the limit exists, then it means that

$$\frac{df(x)}{dx} = \frac{f(x+h) - f(x-h)}{2h} + O(h) \qquad (4.10)$$

where $O(h)$ indicates a correction that, at most, is proportional to h.

The three definitions are equivalent for functions that are differentiable in
x, and the latter is preferable because it is more symmetric.

Notice that even more definitions are possible as long as they agree in the
limit $h \to 0$. Definitions that converge faster as h goes to zero are referred
to as "improvement."

We can easily implement the concept of a numerical derivative in code
by creating a *functional* D that takes a function f and returns the function
$\frac{df(x)}{dx}$ (a functional is a function that returns another function):

Listing 4.1: in file: nlib.py

```
def D(f,h=1e-6): # first derivative of f
    return lambda x,f=f,h=h: (f(x+h)-f(x-h))/2/h
```

We can do the same with the second derivative:

$$\frac{d^2 f(x)}{dx^2} = \frac{f(x+h) - 2f(x) - f(x-h)}{h^2} + O(h) \tag{4.11}$$

Listing 4.2: in file: nlib.py

```
def DD(f,h=1e-6): # second derivative of f
    return lambda x,f=f,h=h: (f(x+h)-2.0*f(x)+f(x-h))/(h*h)
```

Here is an example:

Listing 4.3: in file: nlib.py

```
>>> def f(x): return x*x-5.0*x
>>> print f(0)
0.0
>>> f1 = D(f) # first derivative
>>> print f1(0)
-5.0
>>> f2 = DD(f) # second derivative
>>> print f2(0)
2.00000...
>>> f2 = D(f1) # second derivative
>>> print f2(0)
1.99999...
```

Notice how composing the first derivative twice or computing the second
derivative directly yields a similar result.

We could easily derive formulas for higher-order derivatives and implement them, but they are rarely needed.

4.3.3   Replace nonlinear with linear

Suppose we are interested in the values of $f(x) = \sin(x)$ for values of x between 0 and 0.1:

```
>>> from math import sin
>>> points = [0.01*i for i in xrange(0,11)]
>>> for x in points:
...       print x, sin(x), "%.2f" % (abs(x-sin(x))/sin(x)*100)
0.01 0.009999833... 0.00
0.02 0.019998666... 0.01
0.03 0.029995500... 0.02
0.04 0.039989334... 0.03
0.05 0.049979169... 0.04
0.06 0.059964006... 0.06
0.07 0.069942847... 0.08
0.08 0.079914693... 0.11
0.09 0.089878549... 0.14
0.1 0.0998334166... 0.17
```

Here the first column is the value of x, the second column is the corresponding $\sin(x)$, and the third column is the relative difference (in percent) between x and $\sin(x)$. The difference is always less than 20%; therefore, if we are happy with this precision, then we can replace $\sin(x)$ with x.

This works because any function $f(x)$ can be expanded using a Taylor series. The first order of the Taylor expansion is linear. For values of x sufficiently close to the expansion point, the function can therefore be approximated with its Taylor expansion.

Expanding on the previous example, consider the following code:

```
>>> from math import sin
>>> points = [0.01*i for i in xrange(0,11)]
>>> for x in points:
...       s = x - x*x*x/6
...       print x, math.sin(x), s, ``%.6f'' % (abs(s-sin(x))/(sin(x))*100)
0.01 0.009999833... 0.009999... 0.000000
0.02 0.019998666... 0.019998... 0.000000
0.03 0.029995500... 0.029995... 0.000001
```

```
 9  0.04 0.039989334... 0.039989... 0.000002
10  0.05 0.049979169... 0.049979... 0.000005
11  0.06 0.059964006... 0.059964... 0.000011
12  0.07 0.069942847... 0.069942... 0.000020
13  0.08 0.079914693... 0.079914... 0.000034
14  0.09 0.089878549... 0.089878... 0.000055
15  0.1 0.0998334166... 0.099833... 0.000083
```

Notice that the third column $s = x - x^3/6$ is very close to $\sin(x)$. In fact, the difference is less than one part in 10,000 (fourth column). Therefore, for $x \in [-1, +1]$, it is possible to replace the $sin(x)$ function with the $x - x^3/6$ polynomial. Here we just went one step further in the Taylor expansion, replacing the first order with the third order. The error committed in this approximation is very small.

4.3.4 Transform a problem into a different one

Continuing with the previous example, the polynomial approximation for the sin function works when x is smaller than 1 but fails when x is greater than or equal to 1. In this case, we can use the following relations to reduce the computation of $\sin(x)$ for large x to $\sin(x)$ for $0 < x < 1$. In particular, we can use

$$\sin(x) = -\sin(-x) \text{when} x < 0 \qquad (4.12)$$

to reduce the domain to $x \in [0, \infty]$. We can then use

$$\sin(x) = \sin(x - 2k\pi) \qquad k \in \mathbb{N} \qquad (4.13)$$

to reduce the domain to $x \in [0, 2\pi)$

$$\sin(x) = -\sin(2\pi - x) \qquad (4.14)$$

to reduce the domain to $x \in [0, \pi)$

$$\sin(x) = \sin(\pi - x) \qquad (4.15)$$

to reduce the domain to $x \in [0, \pi/2)$, and

$$\sin(x) = \sqrt{1 - \sin(\pi/2 - x)^2} \qquad (4.16)$$

to reduce the domain to $x \in [0, \pi/4)$, where the latter is a subset of $[0, 1)$.

4.3.5 Approximate the true result via iteration

The approximations $\sin(x) \simeq x$ and $\sin(x) \simeq x - x^3/6$ came from linearizing the function $\sin(x)$ and adding a correction to the previous approximation, respectively. In general, we can iterate the process of finding corrections and approximating the true result.

Here is an example of a general iterative algorithm:

```
1  result=guess
2  loop:
3      compute correction
4      result=result+correction
5      if result sufficiently close to true result:
6          return result
```

For the sin function:

```
1  def mysin(x):
2      (s,t) = (0.0,x)
3      for i in xrange(3,10,2): (s, t) = (s+t, -t*x*x/i/(i-1))
4      return s
```

Where do these formulas come from? How do we decide how many iterations we need? We address these problems in the next section.

4.3.6 Taylor series

A function $f(x) : \mathbb{R} \to \mathbb{R}$ is said to be a *real analytical* in $\bar{x}$ if it is continuous in $x = \bar{x}$ and all its derivatives exist and are continuous in $x = \bar{x}$.

When this is the case, the function can be locally approximated with a local power series:

$$f(x) = f(\bar{x}) + f^{(0)}(\bar{x})(x - \bar{x}) + \dots + \frac{f^{(k)}(\bar{x})}{n!}(x - \bar{x})^k + R_k \qquad (4.17)$$

The remainder R_k can be proven to be (Taylor's theorem):

$$R_k = \frac{f^{(k+1)}(\zeta)}{(k+1)!}(x - \bar{x})^{k+1} \qquad (4.18)$$

where ζ is a point in between x and $\bar{x}$. Therefore, if $f^{(k+1)}$ exists and is limited within a neighborhood $D = \{x \text{ for } |x - \bar{x}| < \epsilon\}$, then

$$|R_k| < \left| max_{x \in D} f^{(k+1)} \right| |(x - \bar{x})^{k+1}| \qquad (4.19)$$

If we stop the Taylor expansion at a finite value of k, the preceding formula gives us the statistical error part of the computational error.

Some Taylor series are very easy to compute:

Exponential for $\bar{x} = 0$:

$$
\begin{aligned}
f(x) &= e^x & (4.20) \\
f^{(1)}(x) &= e^x & (4.21) \\
\dots & \quad \dots & (4.22) \\
f^{(k)}(x) &= e^x & (4.23) \\
e^x &= 1 + x + \frac{1}{2}x^2 + \dots + \frac{1}{k!}x^k + \dots & (4.24)
\end{aligned}
$$

Sin for $\bar{x} = 0$:

$$f(x) = sin(x) \tag{4.25}$$

$$f^{(1)}(x) = cos(x) \tag{4.26}$$

$$f^{(2)}(x) = -sin(x) \tag{4.27}$$

$$f^{(3)}(x) = -cos(x) \tag{4.28}$$

$$\cdots \qquad \cdots \tag{4.29}$$

$$sin(x) = x - \frac{1}{3!}x^3 + \dots + \frac{(-1)^n}{(2k+1)!}x^{(2k+1)} + \dots \tag{4.30}$$

We can show the effects of the various terms:

Listing 4.4: in file: `nlib.py`

```
>>> X = [0.03*i for i in xrange(200)]
>>> c = Canvas(title='sin(x) approximations')
>>> c.plot([(x,math.sin(x)) for x in X],legend='sin(x)')
<...>
>>> c.plot([(x,x) for x in X[:100]],legend='Taylor 1st')
<...>
>>> c.plot([(x,x-x**3/6) for x in X[:100]],legend='Taylor 5th')
<...>
>>> c.plot([(x,x-x**3/6+x**5/120) for x in X[:100]],legend='Taylor 5th')
<...>
>>> c.save('images/sin.png')
```

Notice that we can very well expand in Taylor around any other point, for example, $\bar{x} = \pi/2$, and we get

$$sin(x) = 1 - \frac{1}{2}(x - \frac{\pi}{2})^2 + \dots + \frac{(-1)^n}{(2k)!}(x - \frac{\pi}{2})^{(2k)} + \dots \tag{4.31}$$

and a plot would show:

Listing 4.5: in file: `nlib.py`

```
>>> a = math.pi/2
>>> X = [0.03*i for i in xrange(200)]
>>> c = Canvas(title='sin(x) approximations')
>>> c.plot([(x,math.sin(x)) for x in X],legend='sin(x)')
<...>
>>> c.plot([(x,1-(x-a)**2/2) for x in X[:150]],legend='Taylor 2nd')
<...>
>>> c.plot([(x,1-(x-a)**2/2+(x-a)**4/24) for x in X[:150]], legend='Taylor 4th')
```

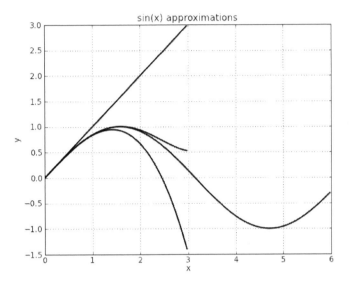

Figure 4.1: The figure shows the sin function and its approximation using the Taylor expansion around $x = 0$ at different orders.

```
9  <...>
10 >>> c.plot([(x,1-(x-a)**2/2+(x-a)**4/24-(x-a)**6/720) for x in X[:150]],legend='
      Taylor 6th')
11 <...>
12 >>> c.save('images/sin2.png')
```

Similarly we can expand the cos function around $\bar{x} = 0$. Not accidentally, we would get the same coefficients as the Taylor expansion of the sin function around $\bar{x} = \pi/2$. In fact, $\sin(x) = \cos(x - \pi/2)$:

$$f(x) \quad = \quad \cos(x) \tag{4.32}$$

$$f^{(1)}(x) \quad = \quad -\sin(x) \tag{4.33}$$

$$f^{(2)}(x) \quad = \quad -\cos(x) \tag{4.34}$$

$$f^{(3)}(x) \quad = \quad \sin(x) \tag{4.35}$$

$$\cdots \qquad \cdots \tag{4.36}$$

$$\cos(x) \quad = \quad 1 - \frac{1}{2}x^2 + \dots + \frac{(-1)^n}{(2k)!}x^{(2k)} + \dots \tag{4.37}$$

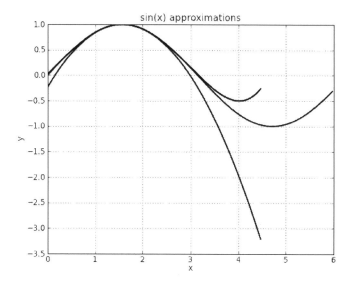

Figure 4.2: The figure shows the sin function and its approximation using the Taylor expansion around $x = \pi/2$ at different orders.

With a simple replacement, it is easy to prove that

$$e^{ix} = \cos(x) + i\sin(x) \tag{4.38}$$

which will be useful when we talk about Fourier and Laplace transforms.

Now let's consider the kth term in Taylor expansion of e^x. It can be rearranged as a function of the previous $(k-1) - th$ term:

$$T_k(x) = \frac{1}{k!}x^n = \frac{x}{k}\frac{1}{(k-1)!}x^{k-1} = \frac{x}{k}T_{k-1}(x) \tag{4.39}$$

For $x < 0$, the terms in the sign have alternating sign and are decreasing in magnitude; therefore, for $x < 0$, $R_k < T_{k+1}(1)$. This allows for an easy implementation of the Taylor expansion and its stopping condition:

Listing 4.6: in file: `nlib.py`

```
 1  def myexp(x,precision=1e-6,max_steps=40):
 2      if x==0:
```

```
3        return 1.0
4    elif x>0:
5        return 1.0/myexp(-x,precision,max_steps)
6    else:
7        t = s = 1.0 # first term
8        for k in xrange(1,max_steps):
9            t = t*x/k   # next term
10           s = s + t   # add next term
11           if abs(t)<precision: return s
12       raise ArithmeticError('no convergence')
```

This code presents all the features of many of the algorithms we see later in the chapter:

- It deals with the special case $e^0 = 1$.

- It reduces difficult problems to easier problems (exponential of a positive number to the exponential of a negative number via $e^x = 1/e^{-x}$).

- It approximates the "true" solution by iterations.

- The max number of iterations is limited.

- There is a stopping condition.

- It detects failure to converge.

Here is a test of its convergence:

Listing 4.7: in file: nlib.py
```
1 >>> for i in xrange(10):
2 ...     x= 0.1*i
3 ...     assert abs(myexp(x) - math.exp(x)) < 1e-4
```

We can do the same for the sin function:

$$T_k(x) = -\frac{x^2}{(2k)(2k + 1)} T_{k-1}(x) \tag{4.40}$$

In this case, the residue is always limited by

$$|R_k| < |x^{2k+1}| \tag{4.41}$$

because the derivatives of sin are always sin and cos and their image is always between $[-1,1]$.

Also notice that the stopping condition is only true when $0 \leq x < 1$.

Therefore, for other values of x, we must use trigonometric relations to reduce the problem to a domain where the Taylor series converges.

Hence we write:

Listing 4.8: in file: nlib.py

```python
def mysin(x,precision=1e-6,max_steps=40):
    pi = math.pi
    if x==0:
        return 0
    elif x<0:
        return -mysin(-x)
    elif x>2.0*pi:
        return mysin(x % (2.0*pi))
    elif x>pi:
        return -mysin(2.0*pi - x)
    elif x>pi/2:
        return mysin(pi-x)
    elif x>pi/4:
        return sqrt(1.0-mysin(pi/2-x)**2)
    else:
        t = s = x                        # first term
        for k in xrange(1,max_steps):
            t = t*(-1.0)*x*x/(2*k)/(2*k+1)  # next term
            s = s + t                       # add next term
            r = x**(2*k+1)                  # estimate residue
            if r<precision: return s       # stopping condition
        raise ArithmeticError('no convergence')
```

Here we test it:

Listing 4.9: in file: nlib.py

```python
>>> for i in xrange(10):
...     x= 0.1*i
...     assert abs(mysin(x) - math.sin(x)) < 1e-4
```

Finally, we can do the same for the cos function:

Listing 4.10: in file: nlib.py

```python
def mycos(x,precision=1e-6,max_steps=40):
    pi = math.pi
    if x==0:
        return 1.0
    elif x<0:
        return mycos(-x)
    elif x>2.0*pi:
        return mycos(x % (2.0*pi))
```

```
 9      elif x>pi:
10          return mycos(2.0*pi - x)
11      elif x>pi/2:
12          return -mycos(pi-x)
13      elif x>pi/4:
14          return sqrt(1.0-mycos(pi/2-x)**2)
15      else:
16          t = s = 1                        # first term
17          for k in xrange(1,max_steps):
18              t = t*(-1.0)*x*x/(2*k)/(2*k-1)    # next term
19              s = s + t                    # add next term
20              r = x**(2*k)                 # estimate residue
21              if r<precision: return s     # stopping condition
22          raise ArithmeticError('no convergence')
```

Here is a test of convergence:

Listing 4.11: in file: `nlib.py`

```
1  >>> for i in xrange(10):
2  ...     x = 0.1*i
3  ...     assert abs(mycos(x) - math.cos(x)) < 1e-4
```

4.3.7 Stopping Conditions

To implement a stopping condition, we have two options. We can look at the absolute error, defined as

$$[\text{absolute error}] = [\text{approximate value}] - [\text{true value}] \qquad (4.42)$$

or we can look at the relative error

$$[\text{relative error}] = [\text{absolute error}]/[\text{true value}] \qquad (4.43)$$

or better, we can consider both. Here is an example of pseudo-code:

```
1  result = guess
2  loop:
3      compute correction
4      result = result+correction
5      compute remainder
6      if |remainder| < target_absolute_precision return result
7      if |remainder| < target_relative_precision*|result| return result
```

In the code, we use the computed `result` as an estimate of the [true value] and, occasionally, the computed `correction` is an estimate of the [absolute error]. The target absolute precision is an input value that we use as an upper limit for the absolute error. The target relative precision is an input value we use as an upper limit for the relative error. When absolute error falls below the target absolute precision or the relative error falls below the target relative precision, we stop looping and assume the result is sufficiently precise:

```
def generic_looping_function(guess, ap, rp, ns):
    result = guess
    for k in xrange(ns):
        correction = ...
        result = result+correction
        remainder = ...
        if norm(remainder) < max(ap,norm(result)*rp): return result
    raise ArithmeticError('no convergence')
```

In the preceding code,

- `ap` is the target absolute precision.

- `rp` is the target relative precision.

- `ns` is the maximum number of steps.

From now on, we will adopt this naming convention.

Consider, for example, a financial algorithm that outputs a dollar amount. If it converges to a number very close to 1 or 0, the concept of relative precision loses significance for a result equal to zero, and the algorithm never detects convergence. In this case, setting an absolute precision of $1 or 1c is the right thing to do. Conversely, if the algorithm converges to a very large dollar amount, setting a precision of $1 or 1c may be a too strong requirement, and the algorithm will take too long to converge. In this case, setting a relative precision of 1% or 0.1% is the correct thing to do.

Because in general we do not know in advance the output of the algorithm, we should use both stopping conditions. We should also detect which of the two conditions causes the algorithm to stop looping and return, so that we can estimate the uncertainty in the result.

4.4   Linear algebra

In this section, we consider the following algorithms:

- Arithmetic operation among matrices

- Gauss–Jordan elimination for computing the inverse of a matrix A

- Cholesky decomposition for factorizing a symmetric positive definite matrix A into LL^t, where L is a lower triangular matrix

- The Jacobi algorithms for finding eigenvalues

- Fitting algorithms based on linear least squares

We will provide examples of applications.

4.4.1   Linear systems

In mathematics, a system described by a function f is linear if it is additive:

$$f(x+y) = f(x) + f(y) \tag{4.44}$$

and if it is homogeneous,

$$f(\alpha x) = \alpha f(x) \tag{4.45}$$

In simpler words, we can say that the output is proportional to the input.

As discussed in the introduction to this chapter, one of the simplest techniques for approximating any unknown system consists of approximating it with a linear system (and this approximation will be correct for some system and not for others).

When we try to model a new system, approximating the system with a linear system is often the first step in describing it in a quantitative way, even if it may turn out that this is not a good approximation.

This is the same as approximating the function f describing the system with the first-order Taylor expansions $f(x+h) - f(x) = f'(x)h$.

For a multidimensional system with input $\mathbf{x}$ (now a vector) and output $\mathbf{y}$ (also a vector, not necessarily of the same size as $\mathbf{x}$), we can still approximate $\mathbf{y} = f(\mathbf{x})$ with $f(\mathbf{y} + \mathbf{h}) - \mathbf{y} \simeq A\mathbf{h}$, yet we need to clarify what this latter equation means.

Given

$$\mathbf{x} \equiv \begin{pmatrix} x_0 \\ x_1 \\ \dots \\ x_{n-1} \end{pmatrix} \qquad \mathbf{y} \equiv \begin{pmatrix} y_0 \\ y_1 \\ \dots \\ y_{m-1} \end{pmatrix} \qquad (4.46)$$

$$A \equiv \begin{pmatrix} a_{00} & a_{01} & \dots & a_{0,n-1} \\ a_{10} & a_{11} & \dots & a_{1,n-1} \\ \dots & \dots & \dots & \dots \\ a_{m-1,0} & a_{m-1,1} & \dots & a_{m-1,n-1} \end{pmatrix} \qquad (4.47)$$

the following equation means

$$\mathbf{y} = f(\mathbf{x}) \simeq A\mathbf{x} \qquad (4.48)$$

which means

$$
\begin{aligned}
y_0 &= f_0((x) &\simeq a_{00}x_0 + a_{01}x_1 + \dots + a_{0,n-1}x_{n-1} && (4.49) \\
y_1 &= f_1((x) &\simeq a_{10}x_0 + a_{11}x_1 + \dots + a_{1,n-1}x_{n-1} && (4.50) \\
y_2 &= f_2((x) &\simeq a_{20}x_0 + a_{21}x_1 + \dots + a_{2,n-1}x_{n-1} && (4.51) \\
\dots &= &\dots && (4.52) \\
y_{m-1} &= f_{m-1}((x) &\simeq a_{m-1,0}x_0 + a_{m-1,1}x_1 + \dots a_{m-1,n-1}x_{n-1} && (4.53)
\end{aligned}
$$

which says that every output variable y_j is approximated with a function proportional to each of the input variables x_i.

A system is linear if the $\simeq$ relations turn out to be exact and can be replaced by $=$ symbols.

As a corollary of the basic properties of a linear system discussed earlier, linear systems have one nice additional property. If we combine two linear systems $y = Ax$ and $z = By$, the combined system is also a linear system $z = (BA)x$.

Elementary algebra is defined as a set of numbers (e.g., real numbers) endowed with the ordinary four elementary operations $(+,-,\times,/)$.

Abstract algebra is a generalization of the concept of elementary algebra to other sets of objects (not necessarily numbers) by definition operations among them such as addition and multiplication.

Linear algebra is the extension of elementary algebra to matrices (and vectors, which can be seen as special types of matrices) by defining the four elementary operations among them.

We will implement them in code using Python. In Python, we can implement a matrix as a list of lists, as follows:

```
>>> A = [[1,2,3],[4,5,6],[7,8,9]]
```

But such an object (a list of lists) does not have the mathematical properties we want, so we have to define them.

First, we define a class representing a matrix:

Listing 4.12: in file: `nlib.py`

```
class Matrix(object):
    def __init__(self,rows,cols=1,fill=0.0):
        """
        Constructor a zero matrix
        Examples:
        A = Matrix([[1,2],[3,4]])
        A = Matrix([1,2,3,4])
        A = Matrix(10,20)
        A = Matrix(10,20,fill=0.0)
        A = Matrix(10,20,fill=lambda r,c: 1.0 if r==c else 0.0)
        """

        if isinstance(rows,list):
            if isinstance(rows[0],list):
                self.rows = [[e for e in row] for row in rows]
            else:
                self.rows = [[e] for e in rows]
        elif isinstance(rows,int) and isinstance(cols,int):
            xrows, xcols = xrange(rows), xrange(cols)
            if callable(fill):
                self.rows = [[fill(r,c) for c in xcols] for r in xrows]
            else:
                self.rows = [[fill for c in xcols] for r in xrows]
        else:
            raise RuntimeError("Unable to build matrix from %s" % repr(rows))
```

```
25    self.nrows = len(self.rows)
26    self.ncols = len(self.rows[0])
```

Notice that the constructor takes the number of rows and columns (cols) of the matrix but also a `fill` value, which can be used to initialize the matrix elements and defaults to zero. It can be callable in case we need to initialize the matrix with row,col dependent values.

The actual matrix elements are stored as a list or array into the `data` member variable. If `optimize=True`, the data are stored in an `array` of double precision floating point numbers ("d"). This optimization will prevent you from building matrices of complex numbers or matrices of arbitrary precision decimal numbers.

Now we define a getter method, a setter method, and a string representation for the matrix elements:

Listing 4.13: in file: `nlib.py`

```
1    def __getitem__(A,coords):
2        " x = A[0,1]"
3        i,j = coords
4        return A.rows[i][j]
5
6    def __setitem__(A,coords,value):
7        " A[0,1] = 3.0 "
8        i,j = coords
9        A.rows[i][j] = value
10
11   def tolist(A):
12       " assert(Matrix([[1,2],[3,4]]).tolist() == [[1,2],[3,4]]) "
13       return A.rows
14
15   def __str__(A):
16       return str(A.rows)
17
18   def flatten(A):
19       " assert(Matrix([[1,2],[3,4]]).flatten() == [1,2,3,4]) "
20       return [A[r,c] for r in xrange(A.nrows) for c in xrange(A.ncols)]
21
22   def reshape(A,n,m):
23       " assert(Matrix([[1,2],[3,4]]).reshape(1,4).tolist() == [[1,2,3,4]]) "
24       if n*m != A.nrows*A.ncols:
25           raise RuntimeError("Impossible reshape")
26       flat = A.flatten()
27       return Matrix(n,m,fill=lambda r,c,m=m,flat=flat: flat[r*m+c])
```

```
28
29    def swap_rows(A,i,j):
30        " assert(Matrix([[1,2],[3,4]]).swap_rows(1,0).tolist() == [[3,4],[1,2]])
          "
31        A.rows[i], A.rows[j] = A.rows[j], A.rows[i]
```

We also define some convenience functions for constructing the identity matrix (given its size) and a diagonal matrix (given the diagonal elements). We make these methods static because they do not act on an existing matrix.

Listing 4.14: in file: `nlib.py`

```
1    @staticmethod
2    def identity(rows=1,e=1.0):
3        return Matrix(rows,rows,lambda r,c,e=e: e if r==c else 0.0)
4
5    @staticmethod
6    def diagonal(d):
7        return Matrix(len(d),len(d),lambda r,c,d=d:d[r] if r==c else 0.0)
```

Now we are ready to define arithmetic operations among matrices. We start with addition and subtraction:

Listing 4.15: in file: `nlib.py`

```
1    def __add__(A,B):
2        """
3        Adds A and B element by element, A and B must have the same size
4        Example
5        >>> A = Matrix([[4,3.0], [2,1.0]])
6        >>> B = Matrix([[1,2.0], [3,4.0]])
7        >>> C = A + B
8        >>> print C
9        [[5, 5.0], [5, 5.0]]
10       """
11       n, m = A.nrows, A.ncols
12       if not isinstance(B,Matrix):
13           if n==m:
14               B = Matrix.identity(n,B)
15           elif n==1 or m==1:
16               B = Matrix([[B for c in xrange(m)] for r in xrange(n)])
17       if B.nrows!=n or B.ncols!=m:
18           raise ArithmeticError('incompatible dimensions')
19       C = Matrix(n,m)
20       for r in xrange(n):
21           for c in xrange(m):
22               C[r,c] = A[r,c]+B[r,c]
23       return C
```

```
24
25      def __sub__(A,B):
26          """
27          Adds A and B element by element, A and B must have the same size
28          Example
29          >>> A = Matrix([[4.0,3.0], [2.0,1.0]])
30          >>> B = Matrix([[1.0,2.0], [3.0,4.0]])
31          >>> C = A - B
32          >>> print C
33          [[3.0, 1.0], [-1.0, -3.0]]
34          """
35          n, m = A.nrows, A.ncols
36          if not isinstance(B,Matrix):
37              if n==m:
38                  B = Matrix.identity(n,B)
39              elif n==1 or m==1:
40                  B = Matrix(n,m,fill=B)
41          if B.nrows!=n or B.ncols!=m:
42              raise ArithmeticError('Incompatible dimensions')
43          C = Matrix(n,m)
44          for r in xrange(n):
45              for c in xrange(m):
46                  C[r,c] = A[r,c]-B[r,c]
47          return C
48      def __radd__(A,B): #B+A
49          return A+B
50      def __rsub__(A,B): #B-A
51          return (-A)+B
52      def __neg__(A):
53          return Matrix(A.nrows,A.ncols,fill=lambda r,c:-A[r,c])
```

With the preceding definitions, we can add matrices to matrices, subtract matrices from matrices, but also add and subtract scalars to and from matrices and vectors (scalars are interpreted as diagonal matrices when added to square matrices and as constant vectors when added to vectors).

Here are some examples:

Listing 4.16: in file: nlib.py

```
1 >>> A = Matrix([[1.0,2.0],[3.0,4.0]])
2 >>> print A + A        # calls A.__add__(A)
3 [[2.0, 4.0], [6.0, 8.0]]
4 >>> print A + 2        # calls A.__add__(2)
5 [[3.0, 2.0], [3.0, 6.0]]
6 >>> print A - 1        # calls A.__add__(1)
7 [[0.0, 2.0], [3.0, 3.0]]
8 >>> print -A           # calls A.__neg__()
```

```
 9  [[-1.0, -2.0], [-3.0, -4.0]]
10  >>> print 5 - A        # calls A.__rsub__(5)
11  [[4.0, -2.0], [-3.0, 1.0]]
12  >>> b = Matrix([[1.0],[2.0],[3.0]])
13  >>> print b + 2        # calls b.__add__(2)
14  [[3.0], [4.0], [5.0]]
```

The class Matrix works with complex numbers as well:

Listing 4.17: in file: nlib.py

```
1  >>> A = Matrix([[1,2],[3,4]])
2  >>> print A + 1j
3  [[(1+1j), (2+0j)], [(3+0j), (4+1j)]]
```

Now we implement multiplication. We are interested in three types of multiplication: multiplication of a scalar by a matrix (__rmul__), multiplication of a matrix by a matrix (__mul__), and scalar product between two vectors (also handled by __mul__):

Listing 4.18: in file: nlib.py

```
 1  def __rmul__(A,x):
 2      "multiplies a number of matrix A by a scalar number x"
 3      import copy
 4      M = copy.deepcopy(A)
 5      for r in xrange(M.nrows):
 6          for c in xrange(M.ncols):
 7              M[r,c] *= x
 8      return M
 9
10  def __mul__(A,B):
11      "multiplies a number of matrix A by another matrix B"
12      if isinstance(B,(list,tuple)):
13          return (A*Matrix(len(B),1,fill=lambda r,c:B[r])).nrows
14      elif not isinstance(B,Matrix):
15          return B*A
16      elif A.ncols == 1 and B.ncols==1 and A.nrows == B.nrows:
17          # try a scalar product ;-)
18          return sum(A[r,0]*B[r,0] for r in xrange(A.nrows))
19      elif A.ncols!=B.nrows:
20          raise ArithmeticError('Incompatible dimension')
21      M = Matrix(A.nrows,B.ncols)
22      for r in xrange(A.nrows):
23          for c in xrange(B.ncols):
24              for k in xrange(A.ncols):
25                  M[r,c] += A[r,k]*B[k,c]
26      return M
```

This allows us the following operations:

Listing 4.19: in file: nlib.py

```
1  >>> A = Matrix([[1.0,2.0],[3.0,4.0]])
2  >>> print(2*A)         # scalar * matrix
3  [[2.0, 4.0], [6.0, 8.0]]
4  >>> print(A*A)         # matrix * matrix
5  [[7.0, 10.0], [15.0, 22.0]]
6  >>> b = Matrix([[1],[2],[3]])
7  >>> print(b*b)         # scalar product
8  14
```

4.4.2 Examples of linear transformations

In this section, we try to provide an intuitive understanding of two-dimensional linear transformations.

In the following code, we consider an image (a set of points) containing a circle and two orthogonal axes. We then apply the following linear transformations to it:

- A_1, which scales the X-axis

- A_2, which scales the Y-axis

- S, which scales both axes

- B_1, which scales the X-axis and then rotates (R) the image 0.5 rad

- B_2, which is neither a scaling nor a rotation; as it can be seen from the image, it does not preserve angles

Listing 4.20: in file: nlib.py

```
1  >>> points = [(math.cos(0.0628*t),math.sin(0.0628*t)) for t in xrange(200)]
2  >>> points += [(0.02*t,0) for t in xrange(50)]
3  >>> points += [(0,0.02*t) for t in xrange(50)]
4  >>> Canvas(title='Linear Transformation',xlab='x',ylab='y',
5  ...        xrange=(-1,1), yrange=(-1,1)).ellipses(points).save('la1.png')
6  >>> def f(A,points,filename):
7  ...     data = [(A[0,0]*x+A[0,1]*y,A[1,0]*x+A[1,1]*y) for (x,y) in points]
8  ...     Canvas(title='Linear Transformation',xlab='x',ylab='y'
9  ...            ).ellipses(points).ellipses(data).save(filename)
10 >>> A1 = Matrix([[0.2,0],[0,1]])
11 >>> f(A1, points, 'la2.png')
```

```
12 >>> A2 = Matrix([[1,0],[0,0.2]])
13 >>> f(A2, points, 'la3.png')
14 >>> S = Matrix([[0.3,0],[0,0.3]])
15 >>> f(S, points, 'la4.png')
16 >>> s, c = math.sin(0.5), math.cos(0.5)
17 >>> R = Matrix([[c,-s],[s,c]])
18 >>> B1 = R*A1
19 >>> f(B1, points, 'la5.png')
20 >>> B2 = Matrix([[0.2,0.4],[0.5,0.3]])
21 >>> f(B2, points, 'la6.png')
```

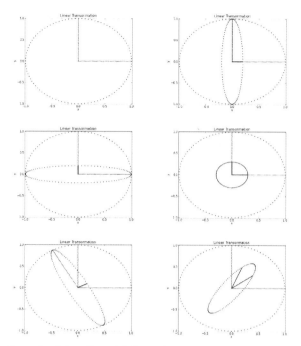

Figure 4.3: Example of the effect of different linear transformations on the same set of points. From left to right, top to bottom, they show stretching along both the X- and Y-axes, scaling across both axes, a rotation, and a generic transformation that does not preserve angles.

4.4.3   Matrix inversion and the Gauss–Jordan algorithm

Implementing the inverse of the multiplication (division) is a more challenging task.

We define A^{-1}, the inverse of the square matrix A, as that matrix such that for every vector b, $A(x) = \mathbf{b}$ implies $(x) = A^{-1}\mathbf{b}$. The Gauss–Jordan algorithm computes A^{-1} given A.

To implement it, we must first understand how it works. Consider the following equation:

$$A\mathbf{x} = \mathbf{b} \qquad (4.54)$$

We can rewrite it as:

$$Ax = Bb \qquad (4.55)$$

where $B = 1$, the identity matrix. This equation remains true if we multiply both terms by a nonsingular matrix S_0:

$$S_0 A x = S_0 B b \qquad (4.56)$$

The trick of the Gauss–Jordan elimination consists in finding a series of matrices $S_0, S_1,..., S_{n-1}$ so that

$$S_{n-1}...S_1 S_0 A x = S_{n-1}...S_1 S_0 B b = x \qquad (4.57)$$

Because the preceding expression must be true for every b and because x is the solution of $Ax = b$, by definition, $S_{n-1} \ldots S_1 S_0 B \equiv A^{-1}$.

The Gauss-Jordan algorithm works exactly this way: given A, it computes A^{-1}:

Listing 4.21: in file: `nlib.py`

```
def __rdiv__(A,x):
    """Computes x/A using Gauss-Jordan elimination where x is a scalar"""
    import copy
    n = A.ncols
    if A.nrows != n:
        raise ArithmeticError('matrix not squared')
    indexes = xrange(n)
    A = copy.deepcopy(A)
    B = Matrix.identity(n,x)
    for c in indexes:
        for r in xrange(c+1,n):
            if abs(A[r,c])>abs(A[c,c]):
                A.swap_rows(r,c)
                B.swap_rows(r,c)
        p = 0.0 + A[c,c] # trick to make sure it is not integer
        for k in indexes:
            A[c,k] = A[c,k]/p
            B[c,k] = B[c,k]/p
        for r in range(0,c)+range(c+1,n):
            p = 0.0 + A[r,c] # trick to make sure it is not integer
            for k in indexes:
                A[r,k] -= A[c,k]*p
                B[r,k] -= B[c,k]*p
        # if DEBUG: print A, B
    return B
```

```
26
27    def __div__(A,B):
28        if isinstance(B,Matrix):
29            return A*(1.0/B) # matrix/matrix
30        else:
31            return (1.0/B)*A # matrix/scalar
```

Here is an example, and we will see many more applications later:

Listing 4.22: in file: nlib.py

```
1  >>> A = Matrix([[1,2],[4,9]])
2  >>> print 1/A
3  [[9.0, -2.0], [-4.0, 1.0]]
4  >>> print A/A
5  [[1.0, 0.0], [0.0, 1.0]]
6  >>> print A/2
7  [[0.5, 1.0], [2.0, 4.5]]
```

4.4.4 Transposing a matrix

Another operation that we will need is transposition:

Listing 4.23: in file: nlib.py

```
1    @property
2    def T(A):
3        """Transposed of A"""
4        return Matrix(A.ncols,A.nrows, fill=lambda r,c: A[c,r])
```

Notice the new matrix is defined with the number of rows and columns switched from matrix A. Notice that in Python, a property is a method that is called like an attribute, therefore without the () notation. This can be used as follows:

Listing 4.24: in file: nlib.py

```
1  >>> A = Matrix([[1,2],[3,4]])
2  >>> print A.T
3  [[1, 3], [2, 4]]
```

For later use, we define two functions to check whether a matrix is symmetrical or zero within a given precision.

Another typical transformation for matrices of complex numbers is the Hermitian operation, which is a transposition combined with complex conjugation of the elements:

Listing 4.25: in file: `nlib.py`

```
@property
def H(A):
    """Hermitian of A"""
    return Matrix(A.ncols,A.nrows, fill=lambda r,c: A[c,r].conj())
```

In later algorithms we will need to check whether a matrix is symmetrical (or almost symmetrical given precision) or zero (or almost zero):

Listing 4.26: in file: `nlib.py`

```
def is_almost_symmetric(A, ap=1e-6, rp=1e-4):
    if A.nrows != A.ncols: return False
    for r in xrange(A.nrows):
        for c in xrange(r):
            delta = abs(A[r,c]-A[c,r])
            if delta>ap and delta>max(abs(A[r,c]),abs(A[c,r]))*rp:
                return False
    return True

def is_almost_zero(A, ap=1e-6, rp=1e-4):
    for r in xrange(A.nrows):
        for c in xrange(A.ncols):
            delta = abs(A[r,c]-A[c,r])
            if delta>ap and delta>max(abs(A[r,c]),abs(A[c,r]))*rp:
                return False
    return True
```

4.4.5 Solving systems of linear equations

Linear algebra is fundamental for solving systems of linear equations such as the following:

$$x_0 + 2x_1 + 2x_2 = 3 \qquad (4.58)$$

$$4x_0 + 4x_1 + 2x_2 = 6 \qquad (4.59)$$

$$4x_0 + 6x_1 + 4x_2 = 10 \qquad (4.60)$$

This can be rewritten using the equivalent linear algebra notation:

$$Ax = b \qquad (4.61)$$

where

$$A = \begin{pmatrix} 1 & 2 & 2 \\ 4 & 4 & 2 \\ 4 & 6 & 4 \end{pmatrix} \quad and \quad b = \begin{pmatrix} 3 \\ 6 \\ 10 \end{pmatrix} \tag{4.62}$$

The solution of the equation can now be written as

$$x = A^{-1}b \tag{4.63}$$

We can easily solve the system with our Python library:

Listing 4.27: in file: nlib.py

```
>>> A = Matrix([[1,2,2],[4,4,2],[4,6,4]])
>>> b = Matrix([[3],[6],[10]])
>>> x = (1/A)*b
>>> print x
[[-1.0], [3.0], [-1.0]]
```

Notice that b is a column vector and therefore

```
>>> b = Matrix([[3],[6],[10]])
```

but not

```
>>> b = Matrix([[3,6,10]]) # wrong
```

We can also obtain a column vector by performing a transposition of a row vector:

```
>>> b = Matrix([[3,6,10]]).T
```

4.4.6 Norm and condition number again

By norm of a vector, we often refer to the 2-norm defined using the Pythagoras theorem:

$$||x||_2 = \sqrt{\sum_i x_i^2} \tag{4.64}$$

For a vector, we can define the p-norm as a generalization of the 2-norm:

$$||x||_p \equiv \left(\sum_i abs(x_i)^p \right)^{\frac{1}{p}} \tag{4.65}$$

We can extend the notation of a norm to any function that maps a vector into a vector, as follows:

$$||f||_p \equiv \max_x ||f(x)||_p / ||x||_p \qquad (4.66)$$

An immediate application is to functions implemented as linear transformations:

$$||A||_p \equiv \max_x ||Ax||_p / ||x||_p \qquad (4.67)$$

This can be difficult to compute in the general case, but it reduces to a simple formula for the 1-norm:

$$||A||_1 \equiv \max_j \sum_i abs(A_{ij}) \qquad (4.68)$$

The 2-norm is difficult to compute for a matrix, but the 1-norm provides an approximation. It is computed by adding up the magnitude of the elements per each column and finding the maximum sum.

This allows us to define a generic function to compute the norm of lists, matrices/vectors, and scalars:

Listing 4.28: in file: nlib.py

```python
def norm(A,p=1):
    if isinstance(A,(list,tuple)):
        return sum(abs(x)**p for x in A)**(1.0/p)
    elif isinstance(A,Matrix):
        if A.nrows==1 or A.ncols==1:
            return sum(norm(A[r,c])**p \
                for r in xrange(A.nrows) \
                for c in xrange(A.ncols))**(1.0/p)
        elif p==1:
            return max([sum(norm(A[r,c]) \
                for r in xrange(A.nrows)) \
                for c in xrange(A.ncols)])
        else:
            raise NotImplementedError
    else:
        return abs(A)
```

Now we can implement a function that computes the condition number for ordinary functions as well as for linear transformations represented by a matrix:

Listing 4.29: in file: nlib.py

```
def condition_number(f,x=None,h=1e-6):
    if callable(f) and not x is None:
        return D(f,h)(x)*x/f(x)
    elif isinstance(f,Matrix): # if is the Matrix
        return norm(f)*norm(1/f)
    else:
        raise NotImplementedError
```

Here are some examples:

Listing 4.30: in file: nlib.py

```
>>> def f(x): return x*x-5.0*x
>>> print condition_number(f,1)
0.74999...
>>> A = Matrix([[1,2],[3,4]])
>>> print condition_number(A)
21.0
```

Having the norm for matrices also allows us to extend the definition of convergence of a Taylor series to a series of matrices:

Listing 4.31: in file: nlib.py

```
def exp(x,ap=1e-6,rp=1e-4,ns=40):
    if isinstance(x,Matrix):
        t = s = Matrix.identity(x.ncols)
        for k in xrange(1,ns):
            t = t*x/k    # next term
            s = s + t    # add next term
            if norm(t)<max(ap,norm(s)*rp): return s
        raise ArithmeticError('no convergence')
    elif type(x)==type(1j):
        return cmath.exp(x)
    else:
        return math.exp(x)
```

Listing 4.32: in file: nlib.py

```
>>> A = Matrix([[1,2],[3,4]])
>>> print exp(A)
[[51.96..., 74.73...], [112.10..., 164.07...]]
```

4.4.7 Cholesky factorization

A matrix is said to be positive definite if $x^t A x > 0$ for every $x \neq 0$.

If a matrix is symmetric and positive definite, then there exists a lower triangular matrix L such that $A = LL^t$. A lower triangular matrix is a matrix that has zeros above the diagonal elements.

The Cholesky algorithm takes a matrix A as input and returns the matrix L:

Listing 4.33: in file: `nlib.py`

```python
def Cholesky(A):
    import copy, math
    if not is_almost_symmetric(A):
        raise ArithmeticError('not symmetric')
    L = copy.deepcopy(A)
    for k in xrange(L.ncols):
        if L[k,k]<=0:
            raise ArithmeticError('not positive definite')
        p = L[k,k] = math.sqrt(L[k,k])
        for i in xrange(k+1,L.nrows):
            L[i,k] /= p
        for j in xrange(k+1,L.nrows):
            p=float(L[j,k])
            for i in xrange(k+1,L.nrows):
                L[i,j] -= p*L[i,k]
    for i in xrange(L.nrows):
        for j in xrange(i+1,L.ncols):
            L[i,j]=0
    return L
```

Here we provide an example and a check that indeed $A = LL^t$:

Listing 4.34: in file: `nlib.py`

```python
>>> A = Matrix([[4,2,1],[2,9,3],[1,3,16]])
>>> L = Cholesky(A)
>>> print is_almost_zero(A - L*L.T)
True
```

The Cholesky algorithm fails if and only if the input matrix is not symmetric or not positive definite, therefore it can be used to check whether a symmetric matrix is positive definite.

Consider for example a generic covariance matrix A. It is supposed to be positive definite, but sometimes it is not, because it is computed incorrectly by taking different subsets of the data to compute A_{ij}, A_{jk}, and A_{ik}. The Cholesky algorithm provides an algorithm to check whether a matrix is positive definite:

Listing 4.35: in file: nlib.py

```
def is_positive_definite(A):
    if not is_almost_symmetric(A):
        return False
    try:
        Cholesky(A)
        return True
    except Exception:
        return False
```

Another application of the Cholesky is in generating vectors **x** with probability distribution

$$p(\mathbf{x}) \propto \exp\left(-\frac{1}{2}\mathbf{x}^t A^{-1}\mathbf{x}\right) \qquad (4.69)$$

where A is a symmetric and positive definite matrix. In fact, if $A = LL^t$, then

$$p(\mathbf{x}) \propto \exp\left(-\frac{1}{2}(L^{-1}\mathbf{x})^t(L^{-1}\mathbf{x})\right) \qquad (4.70)$$

and with a change of variable $\mathbf{u} = \mathbf{L}^{-1}\mathbf{x}$, we obtain

$$p(\mathbf{x}) \propto \exp\left(-\frac{1}{2}\mathbf{u}^t\mathbf{u}\right) \qquad (4.71)$$

and

$$p(\mathbf{x}) \propto e^{-\frac{1}{2}u_0^2}e^{-\frac{1}{2}u_1^2}e^{-\frac{1}{2}u_2^2}... \qquad (4.72)$$

Therefore the u_i components are Gaussian random variables.

In summary, given a covariance matrix A, we can generate random vectors x or random numbers with the same covariance simply by doing

```
def RandomList(A):
    L = Cholesky(A)
    while True:
        u = Matrix([[random.gauss(0,1)] for c in xrange(L.nrows)])
        yield (L*u).flatten()
```

Here is an example of how to use it:

```
>>> A = Matrix([[1.0,0.1],[0.2,3.0]])
>>> for k, v in enumerate(RandomList(A)):
...     print v
```

The RandomList is a generator. You can iterate over it. The yield keyword
is used like return, except the function will return a generator.

4.4.8 Modern portfolio theory

Modern portfolio theory [34] is an investment approach that tries to maximize return given a fixed risk. Many different metrics have been proposed. One of them is the *Sharpe ratio*.

For a stock or a portfolio with an average return r and risk σ, the Sharpe
ratio is defined as

$$\text{Sharpe}(r,\sigma) \equiv (r - \bar{r})/\sigma \tag{4.73}$$

Here $\bar{r}$ is the current risk-free investment rate. Usually the risk is measured as the standard deviation of its daily (or monthly or yearly) return.

Consider the stock price p_{it} of stock i at time t and its arithmetic daily
return $r_{it} = (p_{i,t+1} - p_{it})/p_{it}$ given a risk-free interest equal to $\bar{r}$.

For each stock, we can compute the average return and average risk (variance of daily returns) and display it in a risk-return plot as we did in
chapter 2.

We can try to build arbitrary portfolios by investing in multiple stocks at
the same time. Modern portfolio theory states that there is a maximum
Sharpe ratio and there is one portfolio that corresponds to it. It is called
the tangency portfolio.

A portfolio is identified by fractions of $1 invested in each stock in the
portfolio. Our goal is to determine the tangent portfolio.

If we assume that daily returns for the stocks are Gaussian, then the solving algorithm is simple.

All we need is to compute the average return for each stock, defined as

$$r_i = 1/T \sum_t r_{it} \qquad (4.74)$$

and the covariance matrix

$$A_{ij} = \frac{1}{T} \sum_t (r_{it} - r_i)(r_{jt} - r_j) \qquad (4.75)$$

Modern portfolio theory tells us that the tangent portfolio is given by

$$\mathbf{x} = A^{-1}(\mathbf{r} - \bar{r}\mathbf{1}) \qquad (4.76)$$

The inputs of the formula are the covariance matrix (A), a vector or arithmetic returns for the assets in the portfolio (r), the risk free rate ($\bar{r}$). The output is a vector (x) whose elements are the percentages to be invested in each asset to obtain a tangency portfolio. Notice that some elements of x can be negative and this corresponds to short position (sell, not buy, the asset).

Here is the algorithm:

Listing 4.36: in file: nlib.py

```
def Markowitz(mu, A, r_free):
    """Assess Markowitz risk/return.
    Example:
    >>> cov = Matrix([[0.04, 0.006,0.02],
    ...               [0.006,0.09, 0.06],
    ...               [0.02, 0.06, 0.16]])
    >>> mu = Matrix([[0.10],[0.12],[0.15]])
    >>> r_free = 0.05
    >>> x, ret, risk = Markowitz(mu, cov, r_free)
    >>> print x
    [0.556634..., 0.275080..., 0.1682847...]
    >>> print ret, risk
    0.113915... 0.186747...
    """
    x = Matrix([[0.0] for r in xrange(A.nrows)])
    x = (1/A)*(mu - r_free)
    x = x/sum(x[r,0] for r in xrange(x.nrows))
    portfolio = [x[r,0] for r in xrange(x.nrows)]
    portfolio_return = mu*x
    portfolio_risk = sqrt(x*(A*x))
    return portfolio, portfolio_return, portfolio_risk
```

184 ANNOTATED ALGORITHMS IN PYTHON

Here is an example. We consider three assets (0,1,2) with the following covariance matrix:

```
>>> cov = Matrix([[0.04, 0.006,0.02],
...               [0.006,0.09, 0.06],
...               [0.02, 0.06, 0.16]])
```

and the following expected returns (arithmetic returns, not log returns, because the former are additive, whereas the latter are not):

```
>>> mu = Matrix([[0.10],[0.12],[0.15]])
```

Given the risk-free interest rate

```
>>> r_free = 0.05
```

we compute the tangent portfolio (highest Sharpe ratio), its return, and its risk with one function call:

```
>>> x, ret, risk = Markowitz(mu, cov, r_free)
>>> print x
[0.5566343042071198, 0.27508090614886727, 0.16828478964401297]
>>> print ret, risk
0.113915857605 0.186747095412
>>> print (ret-r_free).risk
0.34225891152
>>> for r in xrange(3): print (mu[r,0]-r_free)/sqrt(cov[r,r])
0.25
0.233333333333
0.25
```

Investing 55% in asset 0, 27% in asset 1, and 16% in asset 2, the resulting portfolio has an expected return of 11.39% and a risk of 18.67%, which corresponds to a Sharpe ratio of 0.34, much higher than 0.25, 0.23, and 0.23 for the individual assets.

Notice that the tangency portfolio is not the only one with the highest Sharpe ratio (return for unit of risk). In fact, any linear combination of the tangency portfolio with a risk-free asset (putting money in the bank) has the same Sharpe ratio. For any target risk, one can find a linear combination of the risk-free asset and the tangent portfolio that has a better Sharpe ratio than any other possible portfolio comprising the same assets.

If we call α the fraction of the money to invest in the tangency portfolio and $1 - \alpha$ the fraction to keep in the bank at the risk free rate, the resulting combined portfolio has return:

$$\alpha \mathbf{x} \cdot \mathbf{r} + (1 - \alpha)\bar{r} \qquad (4.77)$$

and risk

$$\alpha \sqrt{\mathbf{x}^t A \mathbf{x}} \qquad (4.78)$$

We can determine α by deciding how much risk we are willing to take, and these formulas tell us the optimal portfolio for that amount of risk.

4.4.9 Linear least squares, χ^2

Consider a set of data points $(x_j, y_j) = (t_j, o_j \pm do_j)$. We want to fit them with a linear combination of linear independent functions f_i so that

$$
\begin{aligned}
c_0 f_0(t_0) + c_1 f_1(t_0) + c_2 f_2(t_0) + ... &= e_0 \simeq o_0 \pm do_0 & (4.79) \\
c_0 f_0(t_1) + c_1 f_1(t_1) + c_2 f_2(t_1) + ... &= e_1 \simeq o_1 \pm do_1 & (4.80) \\
c_0 f_0(t_2) + c_1 f_1(t_2) + c_2 f_2(t_2) + ... &= e_2 \simeq o_2 \pm do_2 & (4.81) \\
... &= ... & (4.82)
\end{aligned}
$$

We want to find the $\{c_i\}$ that minimizes the sum of the squared distances between the actual "observed" data o_j and the predicted "expected" data e_j, in units of do_j. This metric is called χ^2 in general [35]. An algorithm that minimizes the χ^2 and is linear in the c_i coefficients (our case here) is called *linear least squares* or *linear regression*.

$$\chi^2 = \sum_j \left| \frac{e_j - o_j}{do_j} \right|^2 \qquad (4.83)$$

If we define the matrix A and B as

$$A = \begin{pmatrix} \frac{f_0(t_0)}{do_0} & \frac{f_1(t_0)}{do_0} & \frac{f_2(t_0)}{do_0} & \cdots \\ \frac{f_0(t_1)}{do_1} & \frac{f_1(t_1)}{do_1} & \frac{f_2(t_1)}{do_1} & \cdots \\ \frac{f_0(t_2)}{do_2} & \frac{f_1(t_2)}{do_2} & \frac{f_2(t_2)}{do_2} & \cdots \\ \cdots & \cdots & \cdots & \cdots \end{pmatrix} \qquad b = \begin{pmatrix} \frac{o_0}{do_0} \\ \frac{o_1}{do_1} \\ \frac{o_2}{do_2} \\ \cdots \end{pmatrix} \qquad (4.84)$$

then the problem is reduced to

$$\begin{aligned} \min_{\mathbf{c}} \chi^2 &= \min_{\mathbf{c}} |A\mathbf{c} - \mathbf{b}|^2 & (4.85) \\ &= \min_{\mathbf{c}} (A\mathbf{c} - \mathbf{b})^t (A\mathbf{c} - \mathbf{b}) & (4.86) \\ &= \min_{\mathbf{c}} (\mathbf{c}^t A^t A\mathbf{x} - 2\mathbf{b}^t A\mathbf{c} + \mathbf{b}^t \mathbf{b}) & (4.87) \end{aligned}$$

This is the same as solving the following equation:

$$\begin{aligned} \nabla_c (\mathbf{c}^t A^t A\mathbf{x} - 2\mathbf{c}^t A^t \mathbf{b} + \mathbf{b}^t \mathbf{b}) &= 0 & (4.88) \\ A^t A\mathbf{c} - A^t \mathbf{b} &= 0 & (4.89) \end{aligned}$$

Its solution is

$$\mathbf{c} = (A^t A)^{-1} (A^t \mathbf{b}) \qquad (4.90)$$

The following algorithm implements a fitting function based on the preceding procedure. It takes as input a list of functions f_i and a list of points $p_j = (t_j, o_j, do_j)$ and returns three objects—a list with the c coefficients, the value of χ^2 for the best fit, and the fitting function:

Listing 4.37: in file: `nlib.py`

```
def fit_least_squares(points, f):
    """
    Computes c_j for best linear fit of y[i] \pm dy[i] = fitting_f(x[i])
    where fitting_f(x[i]) is \sum_j c_j f[j](x[i])
```

```
     parameters:
     - a list of fitting functions
     - a list with points (x,y,dy)

     returns:
     - column vector with fitting coefficients
     - the chi2 for the fit
     - the fitting function as a lambda x: ....
     """
     def eval_fitting_function(f,c,x):
         if len(f)==1: return c*f[0](x)
         else: return sum(func(x)*c[i,0] for i,func in enumerate(f))
     A = Matrix(len(points),len(f))
     b = Matrix(len(points))
     for i in xrange(A.nrows):
         weight = 1.0/points[i][2] if len(points[i])>2 else 1.0
         b[i,0] = weight*float(points[i][1])
         for j in xrange(A.ncols):
             A[i,j] = weight*f[j](float(points[i][0]))
     c = (1.0/(A.T*A))*(A.T*b)
     chi = A*c-b
     chi2 = norm(chi,2)**2
     fitting_f = lambda x, c=c, f=f, q=eval_fitting_function: q(f,c,x)
     cs = [c] if isinstance(c,float) else c.flatten()
     return cs, chi2, fitting_f

# examples of fitting functions
def POLYNOMIAL(n):
    return [(lambda x, p=p: x**p) for p in xrange(n+1)]
CONSTANT  = POLYNOMIAL(0)
LINEAR    = POLYNOMIAL(1)
QUADRATIC = POLYNOMIAL(2)
CUBIC     = POLYNOMIAL(3)
QUARTIC   = POLYNOMIAL(4)
```

As an example, we can use it to perform a polynomial fit: given a set of points, we want to find the coefficients of a polynomial that best approximate those points.

In other words, we want to find the c_i such that, given t_j and o_j,

$$c_0 + c_1 t_0^1 + c_2 t_0^2 + \dots \ = \ e_0 \simeq o_0 \pm do_0 \tag{4.91}$$

$$c_0 + c_1 t_1^1 + c_2 t_1^2 + \dots \ = \ e_1 \simeq o_1 \pm do_1 \tag{4.92}$$

$$c_0 + c_1 t_2^1 + c_2 t_2^2 + \dots \ = \ e_2 \simeq o_2 \pm do_2 \tag{4.93}$$

$$\dots \qquad \dots \tag{4.94}$$

$$\tag{4.95}$$

Here is how we can generate some random points and solve the problem for a polynomial of degree 2 (or quadratic fit):

Listing 4.38: in file: `nlib.py`

```
>>> points = [(k,5+0.8*k+0.3*k*k+math.sin(k),2) for k in xrange(100)]
>>> a,chi2,fitting_f = fit_least_squares(points,QUADRATIC)
>>> for p in points[-10:]:
...     print p[0], round(p[1],2), round(fitting_f(p[0]),2)
90 2507.89 2506.98
91 2562.21 2562.08
92 2617.02 2617.78
93 2673.15 2674.08
94 2730.75 2730.98
95 2789.18 2788.48
96 2847.58 2846.58
97 2905.68 2905.28
98 2964.03 2964.58
99 3023.5 3024.48
>>> Canvas(title='polynomial fit',xlab='t',ylab='e(t),o(t)'
...        ).errorbar(points[:10],legend='o(t)'
...        ).plot([(p[0],fitting_f(p[0])) for p in points[:10]],legend='e(t)'
...        ).save('images/polynomialfit.png')
```

Fig. 4.4.9 is a plot of the first 10 points compared with the best fit:

We can also define a $\chi_{dof}^2 = \chi^2 / (N-1)$ where N is the number of c parameters determined by the fit. A value of $\chi_{dof}^2 \simeq 1$ indicates a good fit. In general, the smaller χ_{dof}^2, the better the fit. A large value of χ_{dof}^2 is a symptom of poor modeling (the assumptions of the fit are wrong), whereas a value χ_{dof}^2 much smaller than 1 is a symptom of an overestimate of the errors do_j (or perhaps manufactured data).

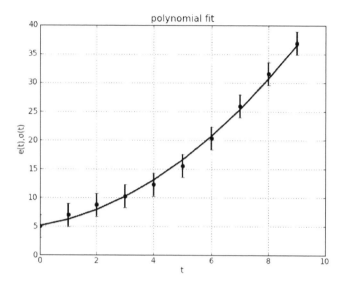

Figure 4.4: Random data with their error bars and the polynomial best fit.

4.4.10   Trading and technical analysis

In finance, *technical analysis* is an empirical discipline that consists of forecasting the direction of prices through the study of patterns in historical data (in particular, price and volume). As an example, we implement a simple strategy that consists of the following steps:

- We fit the adjusted closing price for the previous seven days and use our fitting function to predict the adjusted close for the next day.

- If we have cash and predict the price will go up, we buy the stock.

- If we hold the stock and predict the price will go down, we sell the stock.

Listing 4.39: in file: `nlib.py`

```
class Trader:
    def model(self,window):
        "the forecasting model"
        # we fit last few days quadratically
        points = [(x,y['adjusted_close']) for (x,y) in enumerate(window)]
```

```
6       a,chi2,fitting_f = fit_least_squares(points,QUADRATIC)
7       # and we extrapolate tomorrow's price
8       tomorrow_prediction = fitting_f(len(points))
9       return tomorrow_prediction
10
11   def strategy(self, history, ndays=7):
12       "the trading strategy"
13       if len(history)<ndays:
14           return
15       else:
16           today_close = history[-1]['adjusted_close']
17           tomorrow_prediction = self.model(history[-ndays:])
18           return 'buy' if tomorrow_prediction>today_close else 'sell'
19
20   def simulate(self,data,cash=1000.0,shares=0.0,daily_rate=0.03/360):
21       "find fitting parameters that optimize the trading strategy"
22       for t in xrange(len(data)):
23           suggestion = self.strategy(data[:t])
24           today_close = data[t-1]['adjusted_close']
25           # and we buy or sell based on our strategy
26           if cash>0 and suggestion=='buy':
27               # we keep track of finances
28               shares_bought = int(cash/today_close)
29               shares += shares_bought
30               cash -= shares_bought*today_close
31           elif shares>0 and suggestion=='sell':
32               cash += shares*today_close
33               shares = 0.0
34           # we assume money in the bank also gains an interest
35           cash*=math.exp(daily_rate)
36       # we return the net worth
37       return cash+shares*data[-1]['adjusted_close']
```

Now we back test the strategy using financial data for AAPL for the year 2011:

Listing 4.40: in file: nlib.py

```
1 >>> from datetime import date
2 >>> data = YStock('aapl').historical(
3 ...          start=date(2011,1,1),stop=date(2011,12,31))
4 >>> print Trader().simulate(data,cash=1000.0)
5 1120...
6 >>> print 1000.0*math.exp(0.03)
7 1030...
8 >>> print 1000.0*data[-1]['adjusted_close']/data[0]['adjusted_close']
9 1228...
```

Our strategy did considerably better than the risk-free return of 3% but not as well as investing and holding AAPL shares over the same period.

Of course, we can always engineer a strategy based on historical data that will outperform holding the stock, but *past performance is never a guarantee of future performance.*

According to the definition from investopedia.com, "technical analysts believe that the historical performance of stocks and markets is an indication of future performance."

The efficacy of both technical and fundamental analysis is disputed by the efficient-market hypothesis, which states that stock market prices are essentially unpredictable [36].

It is easy to extend the previous class to implement other strategies and back test them.

4.4.11   Eigenvalues and the Jacobi algorithm

Given a matrix A, an eigenvector is defined as a vector $\mathbf{x}$ such that $A\mathbf{x}$ is proportional to $\mathbf{x}$. The proportionality factor is called an eigenvalue, e. One matrix may have many eigenvectors $\mathbf{x}_i$ and associated eigenvalues e_i:

$$A\mathbf{x}_i = e_i\mathbf{x}_i \tag{4.96}$$

For example:

$$A = \begin{pmatrix} 1 & -2 \\ 1 & 4 \end{pmatrix} \quad and \quad x_i = \begin{pmatrix} -1 \\ 1 \end{pmatrix} \tag{4.97}$$

$$\begin{pmatrix} 1 & -2 \\ 1 & 4 \end{pmatrix} * \begin{pmatrix} -1 \\ 1 \end{pmatrix} = 3 * \begin{pmatrix} -1 \\ 1 \end{pmatrix} \tag{4.98}$$

In this case, x_i is an eigenvector and the corresponding eigenvalue is $e = 3$.

Some eigenvalues may be zero ($e_i = 0$), which means the matrix A is singular. A matrix is singular if it maps a nonzero vector into zero.

Given a square matrix A, if the space generated by the linear indepen-dent eigenvalues has the same dimensionality as the number of rows (or columns) of A, then its eigenvalues are real and the matrix can we written as

$$A = UDU^t \qquad (4.99)$$

where D is a diagonal matrix with eigenvalues on the diagonal $D_{ii} = e_i$ and U is a matrix whose column i is the $\mathbf{x}_i$ eigenvalue.

The following algorithm is called the Jacobi algorithm. It takes as input a symmetric matrix A and returns the matrix U and a list of corresponding eigenvalues e, sorted from smallest to largest:

Listing 4.41: in file: `nlib.py`

```python
def sqrt(x):
    try:
        return math.sqrt(x)
    except ValueError:
        return cmath.sqrt(x)

def Jacobi_eigenvalues(A,checkpoint=False):
    """Returns U end e so that A=U*diagonal(e)*transposed(U)
        where i-column of U contains the eigenvector corresponding to
        the eigenvalue e[i] of A.

        from http://en.wikipedia.org/wiki/Jacobi_eigenvalue_algorithm
    """
    def maxind(M,k):
        j=k+1
        for i in xrange(k+2,M.ncols):
            if abs(M[k,i])>abs(M[k,j]):
                j=i
        return j
    n = A.nrows
    if n!=A.ncols:
        raise ArithmeticError('matrix not squared')
    indexes = xrange(n)
    S = Matrix(n,n, fill=lambda r,c: float(A[r,c]))
    E = Matrix.identity(n)
    state = n
    ind = [maxind(S,k) for k in indexes]
    e = [S[k,k] for k in indexes]
    changed = [True for k in indexes]
    iteration = 0
```

```
31    while state:
32        if checkpoint: checkpoint('rotating vectors (%i) ...' % iteration)
33        m=0
34        for k in xrange(1,n-1):
35            if abs(S[k,ind[k]])>abs(S[m,ind[m]]): m=k
36            pass
37        k,h = m,ind[m]
38        p = S[k,h]
39        y = (e[h]-e[k])/2
40        t = abs(y)+sqrt(p*p+y*y)
41        s = sqrt(p*p+t*t)
42        c = t/s
43        s = p/s
44        t = p*p/t
45        if y<0: s,t = -s,-t
46        S[k,h] = 0
47        y = e[k]
48        e[k] = y-t
49        if changed[k] and y==e[k]:
50            changed[k],state = False,state-1
51        elif (not changed[k]) and y!=e[k]:
52            changed[k],state = True,state+1
53        y = e[h]
54        e[h] = y+t
55        if changed[h] and y==e[h]:
56            changed[h],state = False,state-1
57        elif (not changed[h]) and y!=e[h]:
58            changed[h],state = True,state+1
59        for i in xrange(k):
60            S[i,k],S[i,h] = c*S[i,k]-s*S[i,h],s*S[i,k]+c*S[i,h]
61        for i in xrange(k+1,h):
62            S[k,i],S[i,h] = c*S[k,i]-s*S[i,h],s*S[k,i]+c*S[i,h]
63        for i in xrange(h+1,n):
64            S[k,i],S[h,i] = c*S[k,i]-s*S[h,i],s*S[k,i]+c*S[h,i]
65        for i in indexes:
66            E[k,i],E[h,i] = c*E[k,i]-s*E[h,i],s*E[k,i]+c*E[h,i]
67        ind[k],ind[h]=maxind(S,k),maxind(S,h)
68        iteration+=1
69    # sort vectors
70    for i in xrange(1,n):
71        j=i
72        while j>0 and e[j-1]>e[j]:
73            e[j],e[j-1] = e[j-1],e[j]
74            E.swap_rows(j,j-1)
75            j-=1
76    # normalize vectors
77    U = Matrix(n,n)
78    for i in indexes:
79        norm = sqrt(sum(E[i,j]**2 for j in indexes))
```

```
80        for j in indexes: U[j,i] = E[i,j]/norm
81    return U,e
```

Here is an example that shows, for a particular case, the relation between the input, A, of the output of the U, e of the Jacobi algorithm:

Listing 4.42: in file: `nlib.py`

```
1 >>> import random
2 >>> A = Matrix(4,4)
3 >>> for r in xrange(A.nrows):
4 ...     for c in xrange(r,A.ncols):
5 ...         A[r,c] = A[c,r] = random.gauss(10,10)
6 >>> U,e = Jacobi_eigenvalues(A)
7 >>> print is_almost_zero(U*Matrix.diagonal(e)*U.T-A)
8 True
```

Eigenvalues can be used to filter noise out of data and find hidden dependencies in data. Following are some examples.

4.4.12   Principal component analysis

One important application of the Jacobi algorithm is for principal component analysis (PCA). This is a mathematical procedure that converts a set of observations of possibly correlated vectors into a set of uncorrelated vectors called *principal components*.

Here we consider, as an example, the time series of the adjusted arithmetic returns for the S&P100 stocks that we downloaded and stored in chapter 2.

Each time series is a vector. We know they are not independent because there are correlations. Our goal is to model each time series and a vector plus noise where the vector is the same for all series. We also want find that vector that has maximal superposition with the individual time series, the principal component.

First, we compute the correlation matrix for all the stocks. This is a nontrivial task because we have to make sure that we only consider those days when all stocks were traded:

Listing 4.43: in file: `nlib.py`

```
1   def compute_correlation(stocks, key='arithmetic_return'):
2       "The input must be a list of YStock(...).historical() data"
3       # find trading days common to all stocks
4       days = set()
5       nstocks = len(stocks)
6       iter_stocks = xrange(nstocks)
7       for stock in stocks:
8           if not days: days=set(x['date'] for x in stock)
9           else: days=days.intersection(set(x['date'] for x in stock))
10      n = len(days)
11      v = []
12      # filter out data for the other days
13      for stock in stocks:
14          v.append([x[key] for x in stock if x['date'] in days])
15      # compute mean returns (skip first day, data not reliable)
16      mus = [sum(v[i][k] for k in xrange(1,n))/n for i in iter_stocks]
17      # fill in the covariance matrix
18      var = [sum(v[i][k]**2 for k in xrange(1,n))/n - mus[i]**2 for i in
            iter_stocks]
19      corr = Matrix(nstocks,nstocks,fill=lambda i,j: \
20               (sum(v[i][k]*v[j][k] for k in xrange(1,n))/n - mus[i]*mus[j])/ \
21               math.sqrt(var[i]*var[j]))
22      return corr
```

We use the preceding function to compute the correlation and pass it as input to the Jacobi algorithm and plot the output eigenvalues:

Listing 4.44: in file: nlib.py

```
1   >>> storage = PersistentDictionary('sp100.sqlite')
2   >>> symbols = storage.keys('*/2011')[:20]
3   >>> stocks = [storage[symbol] for symbol in symbols]
4   >>> corr = compute_correlation(stocks)
5   >>> U,e = Jacobi_eigenvalues(corr)
6   >>> Canvas(title='SP100 eigenvalues',xlab='i',ylab='e[i]'
7   ...        ).plot([(i,ei) for i,ei, in enumerate(e)]
8   ...        ).save('images/sp100eigen.png')
```

The image shows that one eigenvalue, the last one, is much larger than the others. It tells us that the data series have something in common. In fact, the arithmetic returns for stock j at time t can be written as

$$r_{it} = \beta_i p_t + \alpha_{it} \tag{4.100}$$

where p is the principal component given by

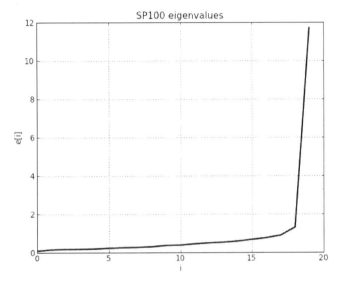

Figure 4.5: Eigenvalues of the correlation matrix for 20 of the S&P100 stocks, sorted by their magnitude.

$$p_t = \sum_i U_{n-1,j} r_{jt} \qquad (4.101)$$

$$\beta_i = \sum_t r_{it} p_t \qquad (4.102)$$

$$\alpha_{it} = r_{it} - \beta_i p_t \qquad (4.103)$$

Here **p** is the vector of adjusted arithmetic returns that better correlates with the returns of the individual assets and therefore best represents the market. The β_i coefficient tells us how much $\mathbf{r}_i$ overlaps with **p**; α, at first approximation, measures leftover noise.

4.5 Sparse matrix inversion

Sometimes we have to invert matrices that are very large, and the Gauss-Jordan algorithms fails. Yet, if the matrix is sparse, in the sense that most of its elements are zeros, than two algorithms come to our rescue: the *minimum residual* and the *biconjugate gradient* (for which we consider a variant called the *stabilized bi-conjugate gradient*).

We will also assume that the matrix to be inverted is given in some implicit algorithmic as $\mathbf{y} = f(\mathbf{x})$ because this is always the case for sparse matrices. There is no point to storing all its elements because most of them are zero.

4.5.1 Minimum residual

Given a linear operator f, the Krylov space spanned by a vector x is defined as

$$K(f,y,i) = \{y, f(y), f(f(y)), f(f(f(y))), (f^i)(y)\} \qquad (4.104)$$

The *minimum residual* [37] algorithm works by solving $x = f^{-1}(y)$ iteratively. At each iteration, it computes a new orthogonal basis vector q_i for the Krylov space $K(f,y,i)$ and computes the coefficients α_i that project x_i into component i of the Krylov space:

$$x_i = y + \alpha_1 q_1 + \alpha_2 q_2 + ... + \alpha_i q_i \in K(f,y,i+1) \qquad (4.105)$$

which minimizes the norm of the residue defined as:

$$r = f(x_i) - y \qquad (4.106)$$

Therefore $\lim_{i \to \infty} f(x_i) = y$. If a solution to the original problem exists, ignoring precision issues, the minimum residual converges to it, and the residue decreases at each iteration.

Notice that in the following code, x and y are exchanged because we adopt the convention that y is the output and x is the input:

Listing 4.45: in file: `nlib.py`

```
def invert_minimum_residual(f,x,ap=1e-4,rp=1e-4,ns=200):
    import copy
    y = copy.copy(x)
    r = x-1.0*f(x)
```

```
5   for k in xrange(ns):
6       q = f(r)
7       alpha = (q*r)/(q*q)
8       y = y + alpha*r
9       r = r - alpha*q
10      residue = sqrt((r*r)/r.nrows)
11      if residue<max(ap,norm(y)*rp): return y
12  raise ArithmeticError('no convergence')
```

4.5.2 Stabilized biconjugate gradient

The stabilized biconjugate gradient [38] method is also based on constructing a Krylov subspace and minimizing the same residue as in the minimum residual algorithm, yet it is faster than the minimum residual and has a smoother convergence than other conjugate gradient methods:

Listing 4.46: in file: nlib.py

```
1   def invert_bicgstab(f,x,ap=1e-4,rp=1e-4,ns=200):
2       import copy
3       y = copy.copy(x)
4       r = x - 1.0*f(x)
5       q = r
6       p = 0.0
7       s = 0.0
8       rho_old = alpha = omega = 1.0
9       for k in xrange(ns):
10          rho = q*r
11          beta = (rho/rho_old)*(alpha/omega)
12          rho_old = rho
13          p = beta*p + r - (beta*omega)*s
14          s = f(p)
15          alpha = rho/(q*s)
16          r = r - alpha*s
17          t = f(r)
18          omega = (t*r)/(t*t)
19          y = y + omega*r + alpha*p
20          residue=sqrt((r*r)/r.nrows)
21          if residue<max(ap,norm(y)*rp): return y
22          r = r - omega*t
23      raise ArithmeticError('no convergence')
```

Notice that the minimum residual and the stabilized biconjugate gradient, if they converge, converge to the same value.

As an example, consider the following. We take a picture using a camera, but we take the picture out of focus. The image is represented by a set of m^2 pixels. The defocusing operation can be modeled as a first approximation with a linear operator acting on the "true" image, x, and turning it into an "out of focus" image, y. We can store the pixels in a one-dimensional vector (both for x and y) as opposed to a matrix by mapping the pixel (r, c) into vector component i using the relation $(r, c) = (i/m, i\%m)$.

Hence we can write

$$\mathbf{y} = A\mathbf{x} \tag{4.107}$$

Here the linear operator A represents the effects of the lens, which transforms one set of pixels into another.

We can model the lens as a sequence of β smearing operators:

$$A = S^{\beta} \tag{4.108}$$

where a smearing operator is a next neighbor interaction among pixels:

$$S_{ij} = (1 - \alpha/4)\delta_{i,j} + \alpha\delta_{i,j\pm1} + \alpha\delta_{i,j\pm m} \tag{4.109}$$

Here α and β are smearing coefficients. When $\alpha = 0$ or $\beta = 0$, the lens has no effect, and $A = I$. The value of α controls how much the value of light at point i is averaged with the value at its four neighbor points: left $(j-1)$, right $(j+1)$, top $(j+m)$, and bottom $(j-m)$. The coefficient β determines the width of the smearing radius. The larger the values of β and α, the more out of focus is the original image.

In the following code, we generate an image x and filter it through a lens operator smear, obtaining a smeared image y. We then use the sparse matrix inverter to reconstruct the original image x given the smeared image y. We use the color2d plotting function to represent the images:

Listing 4.47: in file: `nlib.py`

```
1  >>> m = 30
2  >>> x = Matrix(m*m,1,fill=lambda r,c:(r//m in(10,20) or r%m in(10,20)) and 1. or
       0.)
3  >>> def smear(x):
4  ...     alpha, beta = 0.4, 8
5  ...     for k in xrange(beta):
6  ...         y = Matrix(x.nrows,1)
7  ...         for r in xrange(m):
8  ...             for c in xrange(m):
9  ...                 y[r*m+c,0] = (1.0-alpha/4)*x[r*m+c,0]
10 ...                 if c<m-1: y[r*m+c,0] += alpha * x[r*m+c+1,0]
11 ...                 if c>0:   y[r*m+c,0] += alpha * x[r*m+c-1,0]
12 ...                 if r<m-1: y[r*m+c,0] += alpha * x[r*m+c+m,0]
13 ...                 if c>0:   y[r*m+c,0] += alpha * x[r*m+c-m,0]
14 ...         x = y
15 ...     return y
16 >>> y = smear(x)
17 >>> z = invert_minimum_residual(smear,y,ns=1000)
18 >>> y = y.reshape(m,m)
19 >>> Canvas(title="Defocused image").imshow(y.tolist()).save('images/defocused.
       png')
20 >>> Canvas(title="refocus image").imshow(z.tolist()).save('images/refocused.png'
       )
```

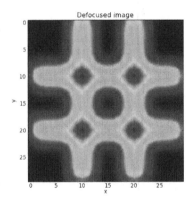
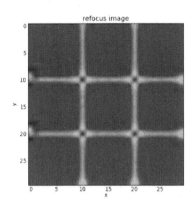

Figure 4.6: An out-of-focus image (left) and the original image (image) computed from the out-of-focus one, using sparse matrix inversion.

When the Hubble telescope was first put into orbit, its mirror was not installed properly and caused the telescope to take pictures out of focus. Until the defect was physically corrected, scientists were able to fix the images using a similar algorithm.

4.6   Solvers for nonlinear equations

In this chapter, we are concerned with the problem of solving in x the equation of one variable:

$$f(x) = 0 \tag{4.110}$$

4.6.1   Fixed-point method

It is always possible to reformulate $f(x) = 0$ as $g(x) = x$ using, for example, one of the following definitions:

- $g(x) = f(x)/c + x$ for some constant c
- $g(x) = f(x)/q(x) + x$ for some $q(x) > 0$ at the solution of $f(x) = 0$

We start at x_0, an arbitrary point in the domain, and close to the solution we seek. We compute

$$
\begin{aligned}
x_1 &= g(x_0) & (4.111)\\
x_2 &= g(x_1) & (4.112)\\
x_3 &= g(x_2) & (4.113)\\
\ldots &= \ldots & (4.114)
\end{aligned}
$$

We can compute the distance between x_i and x as

$$
\begin{aligned}
|x_i - x| &= |g(x_{i-1}) - g(x)| & (4.115)\\
&= |g(x) + g'(\xi)(x_{i-1} - x) - g(x)| & (4.116)\\
&= |g'(\xi)||x_{i-1} - x| & (4.117)
\end{aligned}
$$

where we use *de l'Hopital rule* and ξ is a point in between x and x_{i-1}.

If the magnitude of the first derivative of g, $|g'|$, is less than 1 in a neighborhood of x, and if x_0 is in such a neighborhood, then

$$|x_i - x| = |g'(\xi)||x_{i-1} - x| < |x_{i-1} - x| \tag{4.118}$$

The x_i series will get closer and closer to the solution x.

Here is the process implemented into an algorithm:

Listing 4.48: in file: nlib.py

```
def solve_fixed_point(f, x, ap=1e-6, rp=1e-4, ns=100):
    def g(x): return f(x)+x # f(x)=0 <=> g(x)=x
    Dg = D(g)
    for k in xrange(ns):
        if abs(Dg(x)) >= 1:
            raise ArithmeticError('error D(g)(x)>=1')
        (x_old, x) = (x, g(x))
        if k>2 and norm(x_old-x)<max(ap,norm(x)*rp):
            return x
    raise ArithmeticError('no convergence')
```

And here is an example:

Listing 4.49: in file: nlib.py

```
>>> def f(x): return (x-2)*(x-5)/10
>>> print round(solve_fixed_point(f,1.0,rp=0),4)
2.0
```

4.6.2 Bisection method

The goal of the bisection [39] method is to solve $f(x) = 0$ when the function is continuous and it is known to change sign in between $x = a$ and $x = b$. The bisection method is the continuous equivalent of the binary search algorithm seen in chapter 3. The algorithm iteratively finds the middle point of the domain $x = (b + a)/2$, evaluates the function there, and decides whether the solution is on the left or the right, thus reducing the size of the domain from (a, b) to (a, x) or (x, b), respectively:

Listing 4.50: in file: nlib.py

```
def solve_bisection(f, a, b, ap=1e-6, rp=1e-4, ns=100):
    fa, fb = f(a), f(b)
    if fa == 0: return a
    if fb == 0: return b
    if fa*fb > 0:
        raise ArithmeticError('f(a) and f(b) must have opposite sign')
    for k in xrange(ns):
        x = (a+b)/2
        fx = f(x)
```

```
10      if fx==0 or norm(b-a)<max(ap,norm(x)*rp): return x
11      elif fx * fa < 0: (b,fb) = (x, fx)
12      else: (a,fa) = (x, fx)
13  raise ArithmeticError('no convergence')
```

Here is how to use it:

Listing 4.51: in file: `nlib.py`

```
1  >>> def f(x): return (x-2)*(x-5)
2  >>> print round(solve_bisection(f,1.0,3.0),4)
3  2.0
```

4.6.3 Newton method

The Newton [40] algorithm also solves $f(x) = 0$. It is faster (on average) than the bisection method because it makes the additional assumption that the function is also differentiable. This algorithm starts from an arbitrary point x_0 and approximates the function at that point with its first-order Taylor expansion

$$f(x) \simeq f(x_0) + f'(x_0)(x - x_0) \tag{4.119}$$

and solves it exactly:

$$f(x) = 0 \rightarrow x = x_0 - \frac{f(x_0)}{f'(x_0)} \tag{4.120}$$

thus finding a new and better estimate for the solution. The algorithm iterates the preceding equation, and when it converges, it approximates the exact solution better and better:

Listing 4.52: in file: `nlib.py`

```
1  def solve_newton(f, x, ap=1e-6, rp=1e-4, ns=20):
2      x = float(x) # make sure it is not int
3      for k in xrange(ns):
4          (fx, Dfx) = (f(x), D(f)(x))
5          if norm(Dfx) < ap:
6              raise ArithmeticError('unstable solution')
7          (x_old, x) = (x, x-fx/Dfx)
8          if k>2 and norm(x-x_old)<max(ap,norm(x)*rp): return x
9      raise ArithmeticError('no convergence')
```

The algorithm is guaranteed to converge if $|f'(x)| > 1$ in some neighborhood of the solution and if the starting point is in this neighborhood. It may also converge if this condition is not true. It is likely to fail when $|f'(x)| \simeq 0$ is in the neighborhood of the solution or the starting point because the terms fx/Dfx would become very large.

Here is an example:

Listing 4.53: in file: nlib.py

```
>>> def f(x): return (x-2)*(x-5)
>>> print round(solve_newton(f,1.0),4)
2.0
```

4.6.4 Secant method

The secant method is very similar to the Newton method, except that $f'(x)$ is replaced by a numerical estimate computed using the current point x and the previous point visited by the algorithm:

$$f'(x_i) = \frac{f(x_i) - f(x_{i-1})}{x_i - x_{i-1}} \tag{4.121}$$

$$x_{i+i} = x_i - \frac{f(x_i)}{f'(x_i)} \tag{4.122}$$

As the algorithm approaches the exact solution, the numerical derivative becomes a better and better approximation for the derivative:

Listing 4.54: in file: nlib.py

```
def solve_secant(f, x, ap=1e-6, rp=1e-4, ns=20):
    x = float(x) # make sure it is not int
    (fx, Dfx) = (f(x), D(f)(x))
    for k in xrange(ns):
        if norm(Dfx) < ap:
            raise ArithmeticError('unstable solution')
        (x_old, fx_old,x) = (x, fx, x-fx/Dfx)
        if k>2 and norm(x-x_old)<max(ap,norm(x)*rp): return x
        fx = f(x)
        Dfx = (fx-fx_old)/(x-x_old)
    raise ArithmeticError('no convergence')
```

Here is an example:

Listing 4.55: in file: nlib.py

```
1 >>> def f(x): return (x-2)*(x-5)
2 >>> print round(solve_secant(f,1.0),4)
3 2.0
```

4.7 Optimization in one dimension

While a solver is an algorithm that finds x such that $f(x) = 0$, an optimization algorithm is one that finds the maximum or minimum of the function $f(x)$. If the function is differentiable, this is achieved by solving $f'(x) = 0$.

For this reason, if the function is differentiable twice, we can simply rename all previous solvers and replace $f(x)$ with $f'(x)$ and $f'(x)$ with $f''(x)$.

4.7.1 Bisection method

Listing 4.56: in file: nlib.py

```
1 def optimize_bisection(f, a, b, ap=1e-6, rp=1e-4, ns=100):
2     return solve_bisection(D(f), a, b , ap, rp, ns)
```

Here is an example:

Listing 4.57: in file: nlib.py

```
1 >>> def f(x): return (x-2)*(x-5)
2 >>> print round(optimize_bisection(f,2.0,5.0),4)
3 3.5
```

4.7.2 Newton method

Listing 4.58: in file: nlib.py

```
1 def optimize_newton(f, x, ap=1e-6, rp=1e-4, ns=20):
2     x = float(x) # make sure it is not int
3     (f, Df) = (D(f), DD(f))
4     for k in xrange(ns):
5         (fx, Dfx) = (f(x), Df(x))
6         if Dfx==0: return x
7         if norm(Dfx) < ap:
8             raise ArithmeticError('unstable solution')
```

```
9      (x_old, x) = (x, x-fx/Dfx)
10     if norm(x-x_old)<max(ap,norm(x)*rp): return x
11  raise ArithmeticError('no convergence')
```

Listing 4.59: in file: nlib.py

```
1 >>> def f(x): return (x-2)*(x-5)
2 >>> print round(optimize_newton(f,3.0),3)
3 3.5
```

4.7.3   Secant method

As in the Newton case, the secant method can also be used to find extrema, by replacing f with f':

Listing 4.60: in file: nlib.py

```
1 def optimize_secant(f, x, ap=1e-6, rp=1e-4, ns=100):
2     x = float(x) # make sure it is not int
3     (f, Df) = (D(f), DD(f))
4     (fx, Dfx) = (f(x), Df(x))
5     for k in xrange(ns):
6         if fx==0: return x
7         if norm(Dfx) < ap:
8             raise ArithmeticError('unstable solution')
9         (x_old, fx_old, x) = (x, fx, x-fx/Dfx)
10        if norm(x-x_old)<max(ap,norm(x)*rp): return x
11        fx = f(x)
12        Dfx = (fx - fx_old)/(x-x_old)
13    raise ArithmeticError('no convergence')
```

Listing 4.61: in file: nlib.py

```
1 >>> def f(x): return (x-2)*(x-5)
2 >>> print round(optimize_secant(f,3.0),3)
3 3.5
```

4.7.4   Golden section search

If the function we want to optimize is continuous but not differentiable, then the previous algorithms do not work. In this case, there is one algorithm that comes to our rescue, the golden section [41] search. It is similar to the bisection method, with one caveat; in the bisection method, at each point, we need to know if a function changes sign in between two points, therefore two points are all we need. If instead we are looking for a max or min, we need to know if the function is concave or convex in between those two points. This requires one extra point in between the two. So while the bisection method only needs one point in between $[a, b]$, the golden search needs two points, x_1 and x_2, in between $[a, b]$, and from them it can determine whether the extreme is in $[a, x_2]$ or in $[x_1, b]$. This is also represented pictorially in fig. 4.7.4. The two points are chosen in an optimal way so that at the next iteration, one of the two points can be recycled by leaving the ratio between $x_1 - a$ and $b - x_2$ fixed and equal to 1:

Listing 4.62: in file: nlib.py

```
def optimize_golden_search(f, a, b, ap=1e-6, rp=1e-4, ns=100):
    a,b=float(a),float(b)
    tau = (sqrt(5.0)-1.0)/2.0
    x1, x2 = a+(1.0-tau)*(b-a), a+tau*(b-a)
    fa, f1, f2, fb = f(a), f(x1), f(x2), f(b)
    for k in xrange(ns):
        if f1 > f2:
            a, fa, x1, f1 = x1, f1, x2, f2
            x2 = a+tau*(b-a)
            f2 = f(x2)
        else:
            b, fb, x2, f2 = x2, f2, x1, f1
            x1 = a+(1.0-tau)*(b-a)
            f1 = f(x1)
        if k>2 and norm(b-a)<max(ap,norm(b)*rp): return b
    raise ArithmeticError('no convergence')
```

Here is an example:

Listing 4.63: in file: nlib.py

```
>>> def f(x): return (x-2)*(x-5)
>>> print round(optimize_golden_search(f,2.0,5.0),3)
3.5
```

4.8   Functions of many variables

To be able to work with functions of many variables, we need to introduce the concept of the partial derivative:

$$\frac{\partial f(\mathbf{x})}{\partial x_i} = \lim_{h \to 0} \frac{f(\mathbf{x} + \mathbf{h}_i) - f(\mathbf{x} - \mathbf{h}_i)}{2h} \tag{4.123}$$

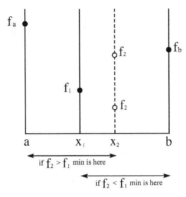

Figure 4.7: Pictorial representation of the golden search method. If the function is concave ($f''(x) > 0$), then knowledge of the function in 4 points (a,x_1,x_2,b) permits us to determine whether a minimum is between $[a, x_2]$ or between $[x_1, b]$.

where $\mathbf{h}_i$ is a vector with components all equal to zero but $h_i = h > 0$.

We can implement it as follows:

Listing 4.64: in file: `nlib.py`

```
def partial(f,i,h=1e-4):
    def df(x,f=f,i=i,h=h):
        x = list(x) # make copy of x
        x[i] += h
        f_plus = f(x)
        x[i] -= 2*h
        f_minus = f(x)
        if isinstance(f_plus,(list,tuple)):
            return [(f_plus[i]-f_minus[i])/(2*h) for i in xrange(len(f_plus))]
        else:
            return (f_plus-f_minus)/(2*h)
    return df
```

Similarly to D(f), we have implemented it in such a way that partial(f,i) returns a function that can be evaluated at any point x. Also notice that the function f may return a scalar, a matrix, a list, or a tuple. The if condition allows the function to deal with the difference between two lists or tuples.

Here is an example:

Listing 4.65: in file: `nlib.py`

```
1  >>> def f(x): return 2.0*x[0]+3.0*x[1]+5.0*x[1]*x[2]
2  >>> df0 = partial(f,0)
3  >>> df1 = partial(f,1)
4  >>> df2 = partial(f,2)
5  >>> x = (1,1,1)
6  >>> print round(df0(x),4), round(df1(x),4), round(df2(x),4)
7  2.0 8.0 5.0
```

4.8.1   Jacobian, gradient, and Hessian

A generic function $f(x_0, x_1, x_2, ...)$ of multiple variables $\mathbf{x} = (x_0, x_1, x_2, ..)$ can be expanded in Taylor series to the second order as

$$
\begin{aligned}
f(x_0, x_1, x_2, ...) \;=\; & f(\bar{x}_0, \bar{x}_1, \bar{x}_2, ...) + & (4.124) \\
& \sum_i \frac{\partial f(\bar{\mathbf{x}})}{\partial x_i}(x_i - \bar{x}_i) + & (4.125) \\
& \sum_{ij} \frac{1}{2}\frac{\partial^2 f}{\partial x_i \partial x_j}(\bar{\mathbf{x}})(x_i - \bar{x}_i)(x_j - \bar{x}_j) + ... & (4.126)
\end{aligned}
$$

We can rewrite the above expression in terms of the vector $\mathbf{x}$ as follows:

$$
f(\mathbf{x}) = f(\bar{\mathbf{x}}) + \nabla_f(\bar{\mathbf{x}})(\mathbf{x} - \bar{\mathbf{x}}) + \frac{1}{2}(\mathbf{x} - \bar{\mathbf{x}})^t H_f(\bar{\mathbf{x}})(\mathbf{x} - \bar{\mathbf{x}}) + ... \qquad (4.127)
$$

where we introduce the gradient vector

$$
\nabla_f(x) \equiv \begin{pmatrix} \partial f(x)/\partial x_0 \\ \partial f(x)/\partial x_1 \\ \partial f(x)/\partial x_2 \\ ... \end{pmatrix} \qquad (4.128)
$$

and the Hessian matrix

$$H_f(x) \equiv \begin{pmatrix} \partial^2 f(x)/\partial x_0 \partial x_0 & \partial^2 f(x)/\partial x_0 \partial x_1 & \partial^2 f(x)/\partial x_0 \partial x_2 & ... \\ \partial^2 f(x)/\partial x_1 \partial x_0 & \partial^2 f(x)/\partial x_1 \partial x_1 & \partial^2 f(x)/\partial x_1 \partial x_2 & ... \\ \partial^2 f(x)/\partial x_2 \partial x_0 & \partial^2 f(x)/\partial x_2 \partial x_1 & \partial^2 f(x)/\partial x_2 \partial x_2 & ... \\ & ... & ... & ... \end{pmatrix}$$

(4.129)

Given the definition of partial, we can compute the gradient and the Hessian using the two functions

Listing 4.66: in file: nlib.py

```
1  def gradient(f, x, h=1e-4):
2      return Matrix(len(x),1,fill=lambda r,c: partial(f,r,h)(x))
3
4  def hessian(f, x, h=1e-4):
5      return Matrix(len(x),len(x),fill=lambda r,c: partial(partial(f,r,h),c,h)(x))
```

Here is an example:

Listing 4.67: in file: nlib.py

```
1  >>> def f(x): return 2.0*x[0]+3.0*x[1]+5.0*x[1]*x[2]
2  >>> print gradient(f, x=(1,1,1))
3  [[1.999999...], [7.999999...], [4.999999...]]
4  >>> print hessian(f, x=(1,1,1))
5  [[0.0, 0.0, 0.0], [0.0, 0.0, 5.000000...], [0.0, 5.000000..., 0.0]]
```

When dealing with functions returning multiple values like

$$f(\mathbf{x}) = (f_0(\mathbf{x}), f_1(\mathbf{x}), f_2(\mathbf{x}), ...)$$

(4.130)

we need to Taylor expand each component:

$$f(\mathbf{x}) = \begin{pmatrix} f_0(\mathbf{x}) \\ f_1(\mathbf{x}) \\ f_2(\mathbf{x}) \\ ... \end{pmatrix} = \begin{pmatrix} f_0(\bar{\mathbf{x}}) + \nabla_{f_0}(\mathbf{x} - \bar{\mathbf{x}}) + ... \\ f_1(\bar{\mathbf{x}}) + \nabla_{f_1}(\mathbf{x} - \bar{\mathbf{x}}) + ... \\ f_2(\bar{\mathbf{x}}) + \nabla_{f_2}(\mathbf{x} - \bar{\mathbf{x}}) + ... \\ ... \end{pmatrix}$$

(4.131)

which we can rewrite as

$$f(\mathbf{x}) = f(\bar{\mathbf{x}}) + J_f(\bar{\mathbf{x}})(\mathbf{x} - \bar{\mathbf{x}}) + ...$$

(4.132)

where J_f is called Jacobian and is defined as

$$J_f \equiv \begin{pmatrix} \partial f_0(x)/\partial x_0 & \partial f_0(x)/\partial x_1 & \partial f_0(x)/\partial x_2 & ... \\ \partial f_1(x)/\partial x_0 & \partial f_1(x)/\partial x_1 & \partial f_1(x)/\partial x_2 & ... \\ \partial f_2(x)/\partial x_0 & \partial f_2(x)/\partial x_1 & \partial f_2(x)/\partial x_2 & ... \\ ... & ... & ... & ... \end{pmatrix} \tag{4.133}$$

which we can implement as follows:

Listing 4.68: in file: nlib.py

```
1  def jacobian(f, x, h=1e-4):
2      partials = [partial(f,c,h)(x) for c in xrange(len(x))]
3      return Matrix(len(partials[0]),len(x),fill=lambda r,c: partials[c][r])
```

Here is an example:

Listing 4.69: in file: nlib.py

```
1  >>> def f(x): return (2.0*x[0]+3.0*x[1]+5.0*x[1]*x[2], 2.0*x[0])
2  >>> print jacobian(f, x=(1,1,1))
3  [[1.9999999..., 7.999999..., 4.9999999...], [1.9999999..., 0.0, 0.0]]
```

4.8.2 Newton method (solver)

We can now solve eq. 4.132 iteratively as we did for the one-dimensional Newton solver with only one change—the first derivative of f is replaced by the Jacobian:

Listing 4.70: in file: nlib.py

```
1  def solve_newton_multi(f, x, ap=1e-6, rp=1e-4, ns=20):
2      """
3      Computes the root of a multidimensional function f near point x.
4
5      Parameters
6      f is a function that takes a list and returns a scalar
7      x is a list
8
9      Returns x, solution of f(x)=0, as a list
10     """
11     n = len(x)
12     x = Matrix(len(x))
13     for k in xrange(ns):
14         fx = Matrix(f(x.flatten()))
```

```
15        J = jacobian(f,x.flatten())
16        if norm(J) < ap:
17            raise ArithmeticError('unstable solution')
18        (x_old, x) = (x, x-(1.0/J)*fx)
19        if k>2 and norm(x-x_old)<max(ap,norm(x)*rp): return x.flatten()
20    raise ArithmeticError('no convergence')
```

Here is an example:

Listing 4.71: in file: nlib.py

```
1 >>> def f(x): return [x[0]+x[1], x[0]+x[1]**2-2]
2 >>> print solve_newton_multi(f, x=(0,0))
3 [1.0..., -1.0...]
```

4.8.3 Newton method (optimize)

As for the one-dimensional case, we can approximate $f(\mathbf{x})$ with its Taylor expansion at the first order,

$$f(\mathbf{x}) = f(\bar{\mathbf{x}}) + \nabla_f(\bar{\mathbf{x}})(\mathbf{x} - \bar{\mathbf{x}}) + \frac{1}{2}(\mathbf{x} - \bar{\mathbf{x}})^t H_f(\bar{\mathbf{x}})(\mathbf{x} - \bar{\mathbf{x}}) \qquad (4.134)$$

set its derivative to zero, and solve it, thus obtaining

$$\mathbf{x} = \bar{\mathbf{x}} - H_f^{-1}\nabla_f \qquad (4.135)$$

which constitutes the core of the multidimensional Newton optimizer:

Listing 4.72: in file: nlib.py

```
1 def optimize_newton_multi(f, x, ap=1e-6, rp=1e-4, ns=20):
2     """
3     Finds the extreme of multidimensional function f near point x.
4
5     Parameters
6     f is a function that takes a list and returns a scalar
7     x is a list
8
9     Returns x, which maximizes of minimizes f(x)=0, as a list
10    """
11    x = Matrix(list(x))
12    for k in xrange(ns):
13        (grad,H) = (gradient(f,x.flatten()), hessian(f,x.flatten()))
14        if norm(H) < ap:
15            raise ArithmeticError('unstable solution')
16        (x_old, x) = (x, x-(1.0/H)*grad)
```

```
17      if k>2 and norm(x-x_old)<max(ap,norm(x)*rp): return x.flatten()
18   raise ArithmeticError('no convergence')
```

Listing 4.73: in file: nlib.py

```
1  >>> def f(x): return (x[0]-2)**2+(x[1]-3)**2
2  >>> print optimize_newton_multi(f, x=(0,0))
3  [2.0, 3.0]
```

4.8.4   Improved Newton method (optimize)

We can further improve the Newton multidimensional optimizer by using the following technique. At each step, if the next guess does not reduce the value of f, we revert to the previous point, and we perform a one-dimensional Newton optimization along the direction of the gradient. This method greatly increases the stability of the multidimensional Newton optimizer:

Listing 4.74: in file: nlib.py

```
1  def optimize_newton_multi_imporved(f, x, ap=1e-6, rp=1e-4, ns=20, h=10.0):
2      """
3      Finds the extreme of multidimensional function f near point x.
4
5      Parameters
6      f is a function that takes a list and returns a scalar
7      x is a list
8
9      Returns x, which maximizes of minimizes f(x)=0, as a list
10     """
11     x = Matrix(list(x))
12     fx = f(x.flatten())
13     for k in xrange(ns):
14         (grad,H) = (gradient(f,x.flatten()), hessian(f,x.flatten()))
15         if norm(H) < ap:
16             raise ArithmeticError('unstable solution')
17         (fx_old, x_old, x) = (fx, x, x-(1.0/H)*grad)
18         fx = f(x.flatten())
19         while fx>fx_old: # revert to steepest descent
20             (fx, x) = (fx_old, x_old)
21             norm_grad = norm(grad)
22             (x_old, x) = (x, x - grad/norm_grad*h)
23             (fx_old, fx) = (fx, f(x.flatten()))
24             h = h/2
25         h = norm(x-x_old)*2
26         if k>2 and h/2<max(ap,norm(x)*rp): return x.flatten()
```

```
27          raise ArithmeticError('no convergence')
```

4.9 Nonlinear fitting

Finally, we have all the ingredients to implement a very generic fitting function that will work linear and nonlinear least squares.

Here we consider a generic experiment or simulated experiment that generates points of the form $(x_i, y_i \pm \delta y_i)$. Our goal is to minimize the χ^2 defined as

$$\chi^2(\mathbf{a}, \mathbf{b}) = \sum_i \left| \frac{y_i - f(x_i, \mathbf{a}, \mathbf{b})}{\delta y_i} \right|^2 \qquad (4.136)$$

where the function f is known but depends on unknown parameters $\mathbf{a} = (a_0, a_1, ...)$ and $\mathbf{b} = (b_0, b_1, ...)$. In terms of these parameters, the function f can be written as follows:

$$f(x, \mathbf{a}, \mathbf{b}) = \sum_j a_j f_j(x, \mathbf{b}) \qquad (4.137)$$

Here is an example:

$$f(x, \mathbf{a}, \mathbf{b}) = a_0 e^{-b_0 x} + a_1 e^{-b_1 x} + a_2 e^{-b_2 x} + ... \qquad (4.138)$$

The goal of our algorithm is to efficiently determine the parameters $\mathbf{a}$ and $\mathbf{b}$ that minimize the χ^2.

We proceed by defining the following two quantities:

$$\mathbf{z} = \begin{pmatrix} y_0 / \delta y_0 \\ y_1 / \delta y_1 \\ y_2 / \delta y_2 \\ ... \end{pmatrix} \qquad (4.139)$$

and

$$A(\mathbf{b}) = \begin{pmatrix} f_0(x_0, \mathbf{b})/\delta y_0 & f_1(x_0, \mathbf{b})/\delta y_0 & f_2(x_0, \mathbf{b})/\delta y_0 & \dots \\ f_0(x_1, \mathbf{b})/\delta y_1 & f_1(x_1, \mathbf{b})/\delta y_1 & f_2(x_1, \mathbf{b})/\delta y_1 & \dots \\ f_0(x_2, \mathbf{b})/\delta y_2 & f_1(x_2, \mathbf{b})/\delta y_2 & f_2(x_2, \mathbf{b})/\delta y_2 & \dots \\ \dots & \dots & \dots & \dots \end{pmatrix} \qquad (4.140)$$

In terms of A and z, the χ^2 can be rewritten as

$$\chi^2(\mathbf{a}, \mathbf{b}) = |A(\mathbf{b})\mathbf{a} - \mathbf{z}|^2 \qquad (4.141)$$

We can minimize this function in a using the linear least squares algorithm, exactly:

$$\mathbf{a}(\mathbf{b}) = (A(\mathbf{b})A(\mathbf{b})^t)^{-1}(A(\mathbf{b})^t z) \qquad (4.142)$$

We define a function that returns the minimum χ^2 for a fixed input $\mathbf{b}$:

$$g(\mathbf{b}) = \min_{\mathbf{a}} \chi^2(\mathbf{a}, \mathbf{b}) = \chi^2(\mathbf{a}(\mathbf{b}), \mathbf{b}) = |A(\mathbf{b})\mathbf{a}(\mathbf{b}) - \mathbf{z}|^2 \qquad (4.143)$$

Therefore we have reduced the original problem to a simple problem by reducing the number of unknown parameters from $N_a + N_b$ to N_b.

The following code takes as input the data as a list of $(x_i, y_i, \delta y_i)$, a list of functions (or a single function), and a guess for the b values. If the fs argument is not a list but a single function, then there is no **a** to compute, and the function proceeds by minimizing the χ^2 using the improved Newton optimizer (the one-dimensional or the improved multidimensional one, as appropriate). If the argument **b** is missing, then the fitting parameters are all linear, and the algorithm reverts to regular linear least squares. Otherwise, run the more complex algorithm described earlier:

Listing 4.75: in file: nlib.py

```
def fit(data, fs, b=None, ap=1e-6, rp=1e-4, ns=200, constraint=None):
    if not isinstance(fs,(list,tuple)):
```

```
3    def g(b, data=data, f=fs, constraint=constraint):
4        chi2 = sum(((y-f(b,x))/dy)**2 for (x,y,dy) in data)
5        if constraint: chi2+=constraint(b)
6        return chi2
7    if isinstance(b,(list,tuple)):
8        b = optimize_newton_multi_imporved(g,b,ap,rp,ns)
9    else:
10        b = optimize_newton(g,b,ap,rp,ns)
11    return b, g(b,data,constraint=None)
12 elif not b:
13    a, chi2, ff = fit_least_squares(data, fs)
14    return a, chi2
15 else:
16    na = len(fs)
17    def core(b,data=data,fs=fs):
18        A = Matrix([[fs[k](b,x)/dy for k in xrange(na)] \
19                                   for (x,y,dy) in data])
20        z = Matrix([[y/dy] for (x,y,dy) in data])
21        a = (1/(A.T*A))*(A.T*z)
22        chi2 = norm(A*a-z)**2
23        return a.flatten(), chi2
24    def g(b,data=data,fs=fs,constraint=constraint):
25        a, chi2 = core(b, data, fs)
26        if constraint:
27            chi += constraint(b)
28        return chi2
29    b = optimize_newton_multi_imporved(g,b,ap,rp,ns)
30    a, chi2 = core(b,data,fs)
31    return a+b,chi2
```

Here is an example:

```
1 >>> data = [(i, i+2.0*i**2+300.0/(i+10), 2.0) for i in xrange(1,10)]
2 >>> fs = [(lambda b,x: x), (lambda b,x: x*x), (lambda b,x: 1.0/(x+b[0]))]
3 >>> ab, chi2 = fit(data,fs,[5])
4 >>> print ab, chi2
5 [0.999..., 2.000..., 300.000..., 10.000...] ...
```

In the preceding implementation, we added a somewhat mysterious argument constraint. This is a function of **b**, and its output gets added to the value of χ^2, which we are minimizing. By choosing the appropriate function, we can set constraints about the expected values b. These constraints represent a priori knowledge about the parameters, that is, knowledge that does not come from the data being fitted.

For example, if we know that b_i must be close to some $\bar{b}_i$ with some uncertainty δb_i, then we can use

```
1  def constraint(b, bar_b, delta_b):
2      return sum(((b[i]-bar_b[i])/delta_b[i])**2 for i in xrange(len(b)))
```

and pass the preceding function as a constraint. From a practical effect, this stabilizes our fit. From a theoretical point of view, the $\bar{b}_i$ are the *priors* of Bayesian statistics.

4.10 Integration

Consider the integral of $f(x)$ for x in domain $[a, b]$, which we normally represent as

$$I = \int_a^b f(x)dx \qquad (4.144)$$

and which measures the area under the curve $y = f(x)$ delimited on the left by $x = a$ and on the right by $x = b$.

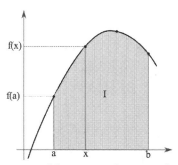

Figure 4.8: Visual representation of the concept of an integral as the area under a curve.

As we did in the previous subsection, we can approximate the possible values taken by x as discrete values $x \equiv hi$, where $h = (b - a)/n$. At those values, the function f evaluates to $f_i \equiv f(hi)$. Thus the integral can be approximated as a sum of trapezoids:

$$I_n \simeq \sum_{i=0}^{i<n} \frac{h}{2}(f_i + f_{i+1}) \qquad (4.145)$$

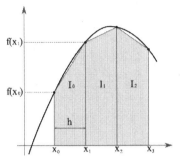

Figure 4.9: Visual representation of the trapezoid method for numerical integration.

If a function is discontinuous only in a finite number of points in the domain $[a, b]$, then the following limit exists:

$$\lim_{n \to \infty} I_n \to I \qquad (4.146)$$

We can implement the naive integration as a function of N as follows:

Listing 4.76: in file: nlib.py

```python
def integrate_naive(f, a, b, n=20):
    """
    Integrates function, f, from a to b using the trapezoidal rule
    >>> from math import sin
    >>> integrate(sin, 0, 2)
    1.416118...
    """
    a,b= float(a),float(b)
    h = (b-a)/n
    return h/2*(f(a)+f(b))+h*sum(f(a+h*i) for i in xrange(1,n))
```

And here we implement the limit by doubling the number of points until convergence is achieved:

Listing 4.77: in file: nlib.py

```python
def integrate(f, a, b, ap=1e-4, rp=1e-4, ns=20):
    """
    Integrates function, f, from a to b using the trapezoidal rule
    converges to precision
    """
    I = integrate_naive(f,a,b,1)
    for k in xrange(1,ns):
```

```
8      I_old, I = I, integrate_naive(f,a,b,2**k)
9      if k>2 and norm(I-I_old)<max(ap,norm(I)*rp): return I
10   raise ArithmeticError('no convergence')
```

We can test the convergence as follows:

Listing 4.78: in file: `nlib.py`

```
1 >>> from math import sin, cos
2 >>> print integrate_naive(sin,0,3,n=2)
3 1.6020...
4 >>> print integrate_naive(sin,0,3,n=4)
5 1.8958...
6 >>> print integrate_naive(sin,0,3,n=8)
7 1.9666...
8 >>> print integrate(sin,0,3)
9 1.9899...
10 >>> print 1.0-cos(3)
11 1.9899...
```

4.10.1 Quadrature

In the previous integration, we divided the domain $[a, b]$ into subdomains, and we computed the area under the curve f in each subdomain by approximating it with a trapezoid; for example, we approximated the function in between x_i and x_{i+1} with a straight line. We can do better by approximating the function with a polynomial of arbitrary degree n and then compute the area in the subdomain by explicitly integrating the polynomial.

This is the basic idea of quadrature. For a subdomain delimited by $(0, 1)$, we can impose

$$\int_0^1 1 dx = h = \sum_i c_i (i/n)^0 \tag{4.147}$$

$$\int_0^1 x dx = h^2/2 = \sum_i c_i (i/n)^1 \tag{4.148}$$

$$\dots \quad \dots \quad \dots \tag{4.149}$$

$$\int_0^1 x^{n-1} dx = h^n/n = \sum_i c_i (i/n)^2 \tag{4.150}$$

where c_i are coefficients to be determined:

Listing 4.79: in file: nlib.py

```
1  class QuadratureIntegrator:
2      """
3      Calculates the integral of the function f from points a to b
4      using n Vandermonde weights and numerical quadrature.
5      """
6      def __init__(self,order=4):
7          h =1.0/(order-1)
8          A = Matrix(order, order, fill = lambda r,c: (c*h)**r)
9          s = Matrix(order, 1, fill = lambda r,c: 1.0/(r+1))
10         w = (1/A)*s
11         self.w = w
12     def integrate(self,f,a,b):
13         w = self.w
14         order = len(w.rows)
15         h = float(b-a)/(order-1)
16         return (b-a)*sum(w[i,0]*f(a+i*h) for i in xrange(order))
17
18 def integrate_quadrature_naive(f,a,b,n=20,order=4):
19     a,b = float(a),float(b)
20     h = float(b-a)/n
21     q = QuadratureIntegrator(order=order)
22     return sum(q.integrate(f,a+i*h,a+i*h+h) for i in xrange(n))
```

Here is an example of usage:

Listing 4.80: in file: nlib.py

```
1  >>> from math import sin
2  >>> print integrate_quadrature_naive(sin,0,3,n=2,order=2)
3  1.60208248595
4  >>> print integrate_quadrature_naive(sin,0,3,n=2,order=3)
5  1.99373945223
6  >>> print integrate_quadrature_naive(sin,0,3,n=2,order=4)
7  1.99164529955
```

4.11 Fourier transforms

A function with a domain over a finite interval $[a,b]$ can be approximated with a vector. For example, consider a function $f(x)$ with domain $[0,T]$. We can sample the function at points $x_k = a + (b-a)k/N$ and represent the discretized function with a vector

$$\mathbf{u}_f \equiv \{cf(x_0), cf(x_1), cf(x_2), ...cf(x_N)\} \qquad (4.151)$$

where c is an arbitrary constant that we choose to be $c = \sqrt{(b-a)/N}$. This choice simplifies our later algebra. Summarizing, we define

$$u_{fk} \equiv \sqrt{\frac{b-a}{N}} f(x_k) \qquad (4.152)$$

Given any two functions, we can define their scalar product as the limit of $N \to \infty$ of the scalar product between their corresponding vectors:

$$f \cdot g \equiv \lim_{N \to \infty} \mathbf{u}_f \cdot \mathbf{u}_g = \lim_{N \to \infty} \frac{b-a}{N} \sum_k f(x_k)g(x_k) \qquad (4.153)$$

Using the definition of integral, it can be proven that, in the limit $N \to \infty$, this is equivalent to

$$f \cdot g = \int_a^b f(x)g(x)dx \qquad (4.154)$$

This is because we have chosen c such that c^2 is the width of a rectangle in the Riemann integration.

From now on, we will omit the f subscript in $\mathbf{u}$ and simply use different letters for vectors representing different sampled functions ($\mathbf{u}$, $\mathbf{v}$, $\mathbf{b}$, etc.).

Because we are interested in numerical algorithms, we will keep N finite and work with the sum instead of the integral.

Given a fixed N, we can always find N vectors $\mathbf{b}_0, \mathbf{b}_1...\mathbf{b}_{N_1}$ that are linearly independent, normalized, and orthogonal, that is,

$$\mathbf{b}_i \cdot \mathbf{b}_j = \sum_k b_{ik} b_{ik} = \delta_{ij} \tag{4.155}$$

Here b_{jk} is the k component of vector $\mathbf{b}_j$ and δ_{ij} is the Kroneker delta defined as 0 when $i \neq j$ and 1 when $i == j$.

Any set of vectors $\{\mathbf{b}_j\}$ meeting the preceding condition is called an orthonormal basis. Any other vector $\mathbf{u}$ can be represented by its projections over the basis vectors:

$$u_i = \sum_i v_j b_{ji} \tag{4.156}$$

where v_j is the projection of u along $\mathbf{b}_j$, which can be computed as

$$v_j = \sum_i u_i b_{ji} \tag{4.157}$$

In fact, by direct substitution, we obtain

$$
\begin{align}
v_j &= \sum_k u_k b_{jk} \tag{4.158} \\
&= \sum_k (\sum_i v_i b_{ik}) b_{jk} \tag{4.159} \\
&= \sum_i v_i (\sum_k b_{ik} b_{jk}) \tag{4.160} \\
&= \sum_i v_i \delta_{ij} \tag{4.161} \\
&= v_j \tag{4.162}
\end{align}
$$

In other words, once we have a basis of vectors, the vector $\mathbf{u}$ can be represented in terms of the vector $\mathbf{v}$ of v_j coefficients and, conversely, $\mathbf{v}$ can be computed from $\mathbf{u}$; $\mathbf{u}$ and $\mathbf{v}$ contain the same information.

The transformation from **u** to **v**, and vice versa, is a linear transformation. We call T^+ the transformation from **u** to **v** and T^- its inverse:

$$\mathbf{v} = T^+(\mathbf{u}) \qquad \mathbf{u} = T^-(\mathbf{v}) \tag{4.163}$$

From the definition, and without attempting any optimization, we can implement these operators as follows:

```
1  def transform(u,b):
2      return [sum(u[k]*bi[k] for k in xrange(len(u))) for bi in b]
3
4  def antitransform(v,b):
5      return [sum(v[i]*bi[k] for i,bi in enumerate(b)) for k in xrange(len(v))]
```

Here is an example of usage:

```
1  >>> def make_basis(N):
2  >>>     return [[1 if i==j else 0 for i in xrange(N)] for j in xrange(N)]
3  >>> b = make_basis(4)
4  >>> print b
5  [[1, 0, 0, 0], [0, 1, 0, 0], [0, 0, 1, 0], [0, 0, 0, 1]]
6  >>> u = [1.0, 2.0, 3.0, 4.0]
7  >>> v = transform(u,b)
8  >>> print antitransform(v,b)
9  [1.0, 2.0, 3.0, 4.0]
```

Of course, this example is trivial because of the choice of basis which makes v the same as u. Yet our argument works for any basis $\mathbf{b}_i$. In particular, we can make the following choice:

$$b_{ji} = \frac{1}{\sqrt{2\pi}} e^{2\pi I i j / N} \tag{4.164}$$

where I is the imaginary unit. With this choice, the T^+ and T^- functions become

$$v_j \underset{FT^+}{=} N^{-\frac{1}{2}} \sum_i u_i e^{2\pi I i j / N} \tag{4.165}$$

$$u_i \underset{FT^-}{=} N^{-\frac{1}{2}} \sum_j v_j e^{-2\pi I i j / N} \tag{4.166}$$

and they take the names of Fourier transform and anti-transform [42], respectively; we can implement them as follows:

```
 1  from cmath import exp as cexp
 2
 3  def fourier(u, sign=1):
 4      N, D = len(u), xrange(len(u))
 5      coeff, omega = 1.0/sqrt(N), 2.0*pi*sign*(1j)/N
 6      return [sum(coeff*u[i]*cexp(omega*i*j) for i in D) for j in D]
 7
 8  def anti_fourier(v):
 9      return fourier(v, sign=-1)
```

Here $1j$ is the Python notation for I and $cexp$ is the exponential function for complex numbers.

Notice how the transformation works even when **u** is a vector of complex numbers.

Something special happens when u is real:

$$Re(v_j) = +Re(v_{N-j-1}) \qquad (4.167)$$
$$Im(v_j) = -Im(v_{N-j-1}) \qquad (4.168)$$

We can speed up the code even more using recursion and by observing that if N is a power of 2, then

$$v_j = N^{-\frac{1}{2}} \sum_i u_{2i} e^{2\pi I(2i)j/N} + \qquad (4.169)$$

$$N^{-\frac{1}{2}} \sum_i u_{2i+1} e^{2\pi I(2i+1)j/N} \qquad (4.170)$$

$$= 2^{-\frac{1}{2}} (v_j^{even} + e^{2\pi j/N} v_j^{even}) \qquad (4.171)$$

where v_j^{even} is the Fourier transform of the even terms and v_j^{odd} is the Fourier transform of the odd terms.

The preceding recursive expression can be implemented using dynamic programming, thus obtaining

```
 1  from cmath import exp as cexp
 2
 3  def fast_fourier(u, sign=1):
```

```
4       N, sqrtN, D = len(u), sqrt(len(u)), xrange(len(u))
5       v = [ui/sqrtN for ui in u]
6       k = N/2
7       while k:
8           omega = cexp(2.0*pi*1j*k/N)
9           for i in D:
10              j = i ^ k
11              if i < k:
12                  ik, jk = int(i/k), int(j/k)
13                  v[i], v[j] = v[i]+(omega**ik)*v[j], v[i]+(omega**jk)*v[j]
14          k/=2
15      return v
16
17  def fast_anti_fourier(v):
18      return fast_fourier(v, sign=-1)
```

This implementation of the Fourier transform is equivalent to the previous one in the sense that it produces the same result (up to numerical issues), but it is faster as it runs in $\Theta(N \log_2 N)$ versus $\Theta(N^2)$ of the naive implementation. Here i ^ j is a binary operator, specifically a XOR. For each binary digit of i, it returns a flipped bit if the corresponding bit in j is 1. For example:

```
1  i  : 10010010101110
2  j  : 00010001000010
3  i^j: 10000011001110
```

4.12 Differential equations

In this section, we deal specifically with differential equations of the following form:

$$a_0 f(x) + a_1 f'(x) + a_2 f''(x) + \dots = g(x) \qquad (4.172)$$

where $f(x)$ is an unknown function to be determined; f', f'', and so on, are its derivatives; a_i are known input coefficients; and $g(x)$ is a known input function:

$$f''(x) - 4f'(x) + f(x) = sin(x) \qquad (4.173)$$

In this case, $a_2(x) = 1$, $a_1(x) = -4$, $a_0(x) = 1$, and $g(x) = \sin(x)$.

This can be solved using Fourier transforms by observing that if the Fourier transform of $f(x)$ is $\tilde{f}(y)$, then the Fourier transform of $f'(x)$ is $iy\tilde{f}(y)$.

Hence, if we Fourier transform both the left and right side of

$$\sum_k a_k f^{(k)}(x) = g(x) \tag{4.174}$$

we obtain

$$\left(\sum_k a_k (iy)^k\right)\tilde{f}(y) = \tilde{g}(y) \tag{4.175}$$

therefore $f(x)$ is the anti-Fourier transform of

$$\tilde{f}(y) = \frac{\tilde{g}(y)}{\left(\sum_k a_k (iy)^k\right)} \tag{4.176}$$

In one equation, the solution of eq. 4.172 is

$$f(x) = T^-(T^+(g)/(\sum_k a_k (iy)^k)) \tag{4.177}$$

This is fine and useful when the Fourier transformations are easy to compute.

A more practical numerical solution is the following. We define

$$y_i(x) \equiv f^{(i)}(x) \tag{4.178}$$

and we rewrite the differential equation as

$$y_0' = y_1 \tag{4.179}$$
$$y_1' = y_2 \tag{4.180}$$
$$y_2' = y_3 \tag{4.181}$$
$$\cdots \qquad \cdots \tag{4.182}$$
$$y_{N-1}' = y_N = (g(x) - \sum_{k<N} a_k y_k(x))/a_N(x) \tag{4.183}$$

or equivalently

$$\mathbf{y}' = F(\mathbf{y}) \tag{4.184}$$

where

$$F(\mathbf{y}) = \mathbf{y} + \begin{pmatrix} y_1 \\ y_2 \\ y_3 \\ \cdots \\ (g(x) - \sum_{k<N} a_k(x) y_k(x))/a_N(x) \end{pmatrix} \tag{4.185}$$

The naive solution is due to Euler:

$$\mathbf{y}(x + h) = \mathbf{y}(x) + hF(\mathbf{y}, x) \tag{4.186}$$

The solution is found by iterating the latest equation. Here h is an arbitrary discretization step. Euler's method works even if the a_k coefficients depend on x.

Although the Euler integrator works in theory, its systematic error adds up and does not disappear in the limit $h \to 0$. More accurate integrators are the Runge–Katta and the Adam–Bashforth. In the fourth-order Runge–Katta, the *classical Runge–Katta method*, we also solve the differential equation by iteration, except that eq. 4.186 is replaced with

$$\mathbf{y}(x + h) = \mathbf{y}(x) + h/6(\mathbf{k}_1 + 2\mathbf{k}_2 + 2\mathbf{k}_3 + \mathbf{k}_4) \tag{4.187}$$

where

$$\mathbf{k}_1 = F(\mathbf{y}, x) \tag{4.188}$$

$$\mathbf{k}_2 = F(\mathbf{y} + h k_1/2, x + h/2) \tag{4.189}$$

$$\mathbf{k}_3 = F(\mathbf{y} + h k_2/2, x + h/2) \tag{4.190}$$

$$\mathbf{k}_4 = F(\mathbf{y} + h k_3, x + h) \tag{4.191}$$

5
Probability and Statistics

5.1 Probability

Probability derives from the Latin *probare* (to prove or to test). The word probably means roughly "likely to occur" in the case of possible future occurrences or "likely to be true" in the case of inferences from evidence. See also probability theory.

What mathematicians call probability is the mathematical theory we use to describe and quantify uncertainty. In a larger context, the word *probability* is used with other concerns in mind. Uncertainty can be due to our ignorance, deliberate mixing or shuffling, or due to the essential randomness of Nature. In any case, we measure the uncertainty of events on a scale from zero (impossible events) to one (certain events or no uncertainty).

There are three standard ways to define probability:

- (frequentist) Given an experiment and a set of possible outcomes S, the probability of an event $A \subset S$ is computed by repeating the experiment N times, counting how many times the event A is realized, N_A, then taking the limit

$$\text{Prob}(A) \equiv \lim_{N \to \infty} \frac{N_A}{N} \qquad (5.1)$$

This definition actually requires that one performs an experiment, if not an infinite, then a number of times.

- (a priori) Given an experiment and a set of possible outcomes S with cardinality $c(S)$, the probability of an event $A \subset S$ is defined as

$$\text{Prob}(A) \equiv \frac{c(A)}{c(S)} \qquad (5.2)$$

This definition is ambiguous because it assumes that each "atomic" event $x \in S$ has the same a priori probability and therefore the definition itself is circular. Nevertheless we use this definition in many practical circumstances. What is the probability that when rolling a dice we will get an even number? The space of possible outcomes is $S = \{1,2,3,4,5,6\}$ and $A = \{2,4,6\}$ therefore $\text{Prob}(A) = c(A)/c(S) = 3/6 = 1/2$. This analysis works for an ideal die and ignores the fact that a real dice may be biased. The former definition takes into account this possibility, whereas the latter does not.

- (axiomatic definition) Given an experiment and a set of possible outcomes S, the probability of an event $A \subset S$ is a number $\text{Prob}(A) \in [0,1]$ that satisfies the following conditions: $\text{Prob}(S) = 1$; $\text{Prob}(A_1 \cup A_2) = \text{Prob}(A_1) + \text{Prob}(A_2)$ if $A_1 \cap A_2 = 0$.

In some sense, probability theory is a physical theory because it applies to the physical world (this is a nontrivial fact). While the axiomatic definition provides the mathematical foundation, the a priori definition provides a method to make predictions based on combinatorics. Finally the *frequentist* definition provides an experimental technique to confront our predictions with experiment (is our dice a perfect dice, or is it biased?).

We will differentiate between an "atomic" event defined as an event that can be realized by a single possible outcome of our experiment and a general event defined as a subset of the space of all possible outcomes. In the case of a dice, each possible number (from 1 to 6) is an event and is also an atomic event. The event of getting an even number is an event but not an atomic event because it can be realized in three possible ways.

The axiomatic definition makes it easy to prove theorems, for example,

If $S = A \cup A^c$ and $A \cap A^c = 0$ then $Prob(A) = 1 - Prob(A^c)$

Python has a module called random that can generate random numbers, and we can use it to perform some experiments. Let's simulate a dice with six possible outcomes. We can use the frequentist definition:

Listing 5.1: in file: `nlib.py`

```
>>> import random
>>> S = [1,2,3,4,5,6]
>>> def Prob(A, S, N=1000):
...         return float(sum(random.choice(S) in A for i in xrange(N)))/N
>>> Prob([6],S)
0.166
>>> Prob([1,2],S)
0.308
```

Here Prob(A) computes the probability that the event is set A using N=1000 simulated experiments. The random.choice function picks one of the choices at random with equal probability.

We can compute the same quantity using the a priori definition:

Listing 5.2: in file: `nlib.py`

```
>>> def Prob(A, S): return float(len(A))/len(S)
>>> Prob([6],S)
0.16666666666666666
>>> Prob([1,2],S)
0.3333333333333333
```

As stated before, the latter is more precise because it produces results for an "ideal" dice while the frequentist's approach produces results for a real dice (in our case, a simulated dice).

5.1.1 Conditional probability and independence

We define $Prob(A|B)$ as the probability of event A given event B, and we write

$$Prob(A|B) \equiv \frac{Prob(AB)}{Prob(B)} \qquad (5.3)$$

where $Prob(AB)$ is the probability that A and B both occur and $Prob(B)$ is the probability that B occurs. Note that if $Prob(A|B) = Prob(A)$, then we say that A and B are independent. From eq.(5.3) we conclude $Prob(AB) =$

$\text{Prob}(A)\text{Prob}(B)$; therefore the probability that two independent events occur is the product of the probability that each individual event occurs.

We can experiment with conditional probability using Python. Let's consider two dices, X and Y. The space of all possible outcomes is given by $S^2 = S \times S$. And we are interested in the probability of the second die giving a 6 given that the first dice is also a 6:

Listing 5.3: in file: nlib.py

```
>>> def cross(u,v): return [(i,j) for i in u for j in v]
>>> def Prob_conditional(A, B, S): return Prob(cross(A,B),cross(S,S))/Prob(B,S)
>>> Prob_conditional([6],[6],S)
0.16666666666666666
```

Because we are only considering cases in which the second die is 6, we will pretend that when the second die is 1 through 5 didn't occur. Not surprisingly, we find that `Prob_conditional([6],[6],S)` produces the same result as `Prob([6],S)` because the two dices are independent.

In fact, we say that two sets of events A and B are independent if and only if $P(A|B) = P(A)$.

5.1.2 Discrete random variables

If S is in the space of all possible outcomes of an experiment and we associate an integer number X to each element of S, we say that X is a *discrete random variable*. If X is a discrete variable, we define $p(x)$, the *probability mass function* or *distribution*, as the probability that $X = x$:

$$p(x) \equiv \text{Prob}(X = x) \tag{5.4}$$

We also define the *expectation value* of any function of a discrete random variable $f(X)$ as

$$E[f(X)] \equiv \sum_i f(x_i)p(x_i) \tag{5.5}$$

where i loops all possible variables x_i of the random variable X.

For example, if X is the random variable associated with the outcome of rolling a dice, $p(x) = 1/6$ if $x = 1,2,3,4,5$ or 6 and $p(x) = 0$ otherwise:

$$E[X] = \sum_i x_i p(x_i) = \sum_{x_i \in \{1,2,3,4,5,6\}} x_i \frac{1}{6} = 3.5 \qquad (5.6)$$

and

$$E[(X - 3.5)^2] = \sum_i (x_i - 3.5)^2 p(x_i) = \sum_{x_i \in \{1,2,3,4,5,6\}} (x_i - 3.5)^2 \frac{1}{6} = 2.9167$$
$$(5.7)$$

We call $E[X]$ the *mean* of X and usually denote it with μ_X. We call $E[(X - \mu_X)^2]$ the *variance* of X and denote it with σ_X^2. Note that

$$\sigma_X = E[X^2] - E[X]^2 \qquad (5.8)$$

For discrete random variables, we can implement these definitions as follows:

Listing 5.4: in file: `nlib.py`

```
def E(f,S): return float(sum(f(x) for x in S))/(len(S) or 1)
def mean(X): return E(lambda x:x, X)
def variance(X): return E(lambda x:x**2, X) - E(lambda x:x, X)**2
def sd(X): return sqrt(variance(X))
```

which we can test with a simulated experiment:

Listing 5.5: in file: `nlib.py`

```
>>> S = [random.random()+random.random() for i in xrange(100)]
>>> print mean(S)
1.000...
>>> print sd(S)
0.4...
```

As another example, let's consider a simple bet on a dice. We roll the dice once and win $20 if the dice returns 6; we lose $5 otherwise:

Listing 5.6: in file: `nlib.py`

```
>>> S = [1,2,3,4,5,6]
>>> def payoff(x): return 20.0 if x==6 else -5.0
>>> print E(payoff,S)
-0.83333...
```

The average expected payoff is −0.83..., which means that on average, we should expect to lose 83 cents at this game.

5.1.3 Continuous random variables

If S is the space of all possible outcomes of an experiment and we associate a real number X with each element of S, we say that X is a *continuous random variable*. We also define a *cumulative distribution function* $F(x)$ as the probability that $X \leq x$:

$$F(x) \equiv \text{Prob}(X \leq x) \qquad (5.9)$$

If S is a continuous set and X is a continuous random variable, then we define a *probability density* or *distribution* $p(x)$ as

$$p(x) \equiv \frac{dF(x)}{dx} \qquad (5.10)$$

and the probability that X falls into an interval $[a, b]$ can be computed as

$$\text{Prob}(a \leq X \leq b) = \int_a^b p(x)dx \qquad (5.11)$$

We also define the *expectation value* of any function of a random variable $f(X)$ as

$$E[f(X)] = \int_{-\infty}^{\infty} f(x)p(x)dx \qquad (5.12)$$

For example, if X is a uniform random variable (probability density $p(x)$ equal to 1 if $x \in [0, 1]$, equal to 0 otherwise)

$$E[X] = \int_{-\infty}^{\infty} xp(x)dx = \int_0^1 x\,dx = \frac{1}{2} \qquad (5.13)$$

and

$$E[(X - \frac{1}{2})^2] = \int_{-\infty}^{\infty} (x - \frac{1}{2})^2 p(x)dx = \int_0^1 (x^2 - x + \frac{1}{4})dx = \frac{1}{12} \qquad (5.14)$$

We call $E[X]$ the **mean** of X and usually denote it with μ_X. We call $E[(X - \mu_X)^2]$ the **variance** of X and denote it with σ_X^2. Note that

$$\sigma_X^2 = E[X^2] - E[X]^2 \qquad (5.15)$$

By definition,

$$F(\infty) \equiv \text{Prob}(X \leq \infty) = 1 \tag{5.16}$$

therefore

$$\text{Prob}(-\infty \leq X \leq \infty) = \int_{-\infty}^{\infty} p(x)dx = 1 \tag{5.17}$$

The distribution p is always normalized to 1.

Moreover,

$$
\begin{aligned}
E[aX + b] &= \int_{-\infty}^{\infty} (ax + b)p(x)dx \tag{5.18} \\
&= a\int_{-\infty}^{\infty} xp(x)dx + b\int_{-\infty}^{\infty} p(x)dx \tag{5.19} \\
&= aE[X] + b \tag{5.20}
\end{aligned}
$$

therefore $E[X]$ is a linear operator.

One important consequence of all these formulas is that if we have a function f and a domain $[a, b]$, we can compute its integral by choosing p to be a uniform distribution with values exclusively between a and b:

$$E[f] = \int_{-\infty}^{\infty} f(x)p(x)dx = \frac{1}{b-a} \int_{a}^{b} f(x)dx \tag{5.21}$$

We can also compute the same integral by using the definition of expectation value for a discrete distribution:

$$E[f] = \sum_{x_i} f(x_i)p(x_i) = \frac{1}{N} \sum_{x_i} f(x_i) \tag{5.22}$$

where x_i are N random points drawn from the uniform distribution p defined earlier. In fact, in the large N limit,

$$\lim_{N \to \infty} \frac{1}{N} \sum_{x_i} f(x_i) = \int_{-\infty}^{\infty} f(x)p(x)dx = \frac{1}{b-a} \int_{a}^{b} f(x)dx \tag{5.23}$$

We can verify the preceding relation numerically for a special case:

Listing 5.7: in file: nlib.py

```
1 >>> from math import sin, pi
2 >>> def integrate_mc(f,a,b,N=1000):
3 ...     return sum(f(random.uniform(a,b)) for i in xrange(N))/N*(b-a)
4 >>> print integrate_mc(sin,0,pi,N=10000)
5 2.000....
```

This is the simplest case of Monte Carlo integration, which is the subject of a following chapter.

5.1.4 Covariance and correlations

Given two random variables, X and Y, we define the covariance (cov) and the correlation (corr) between them as

$$
\begin{aligned}
\text{cov}(X,Y) &\equiv E[(X - \mu_X)(Y - \mu_Y)] = E[XY] - E[X]E[Y] &\quad (5.24)\\
\text{corr}(X,Y) &\equiv \text{cov}(X,Y)/(\sigma_X \sigma_Y) &\quad (5.25)
\end{aligned}
$$

Applying the definitions:

$$
\begin{aligned}
E[XY] &= \int \int xy p(x,y) dx dy &\quad (5.26)\\
&= \int \int xy p(x) p(y) dx dy &\quad (5.27)\\
&= \left[\int x p(x) dx \right] \left[\int y p(y) dy \right] &\quad (5.28)\\
&= E[X]E[Y] &\quad (5.29)
\end{aligned}
$$

therefore

$$
\text{cov}(X,Y) = E[XY] - E[X]E[Y] = 0 \quad (5.30)
$$

Therefore

$$
\sigma^2_{X+Y} = \sigma^2_X + \sigma^2_Y + 2\text{cov}(X,Y) \quad (5.31)
$$

and if X and Y are independent, then $\text{cov}(X,Y) = \text{corr}(X,Y) = 0$.

Notice that the reverse is not true. Even if the correlation and the covariance are zero, X and Y may be dependent.

Moreover,

$$\begin{aligned}
\text{cov}(X,Y) &= E[(X-\mu_X)(Y-\mu_Y)] & (5.32)\\
&= E[(X-\mu_X)(\pm X \mp \mu_X)] & (5.33)\\
&= \pm E[(X-\mu_X)(X-\mu_X)] & (5.34)\\
&= \pm\sigma_X^2 & (5.35)
\end{aligned}$$

Therefore, if X and Y are completely correlated or anti-correlated ($Y = \pm X$), then $\text{cov}(X,Y) = \pm\sigma_X^2$ and $\text{corr}(X,Y) = \pm 1$. Notice that the correlation lies always in the range $[-1,1]$.

Finally, notice that for uncorrelated random variables X_i,

$$\begin{aligned}
E[\sum_i a_i X_i] &= \sum_i a_i E[X_i] & (5.36)\\
E[(\sum_i X_i)^2] &= \sum_i E[X_i^2] & (5.37)
\end{aligned}$$

We can define covariance and correlation for discrete distributions:

Listing 5.8: in file: nlib.py

```
def covariance(X,Y):
    return sum(X[i]*Y[i] for i in xrange(len(X)))/len(X) - mean(X)*mean(Y)
def correlation(X,Y):
    return covariance(X,Y)/sd(X)/sd(Y)
```

Here is an example:

Listing 5.9: in file: nlib.py

```
>>> X = []
>>> Y = []
>>> for i in xrange(1000):
...     u = random.random()
...     X.append(u+random.random())
...     Y.append(u+random.random())
>>> print mean(X)
0.989780352018
>>> print sd(X)
0.413861115381
>>> print mean(Y)
1.00551523013
>>> print sd(Y)
0.404909628555
```

```
15 >>> print covariance(X,Y)
16 0.0802804358268
17 >>> print correlation(X,Y)
18 0.479067813484
```

5.1.5 Strong law of large numbers

If $X_1, X_2, ... X_n$ are a sequence of independent and identically distributed random variables with $E[X_i] = \mu$ and finite variance, then

$$\lim_{n \to \infty} \frac{X_1 + X_2 + ... + X_n}{n} = \mu \qquad (5.38)$$

This theorem means that "the average of the results obtained from a large number of trials should be close to the expected value, and will tend to become closer as more trials are performed." The name of this law is due to Poisson [43].

5.1.6 Central limit theorem

This is one of the most important theorems concerning distributions [44]: if $X_1, X_2, ... X_n$ are a sequence of random variables with finite means, μ_i, and finite variance, σ_i^2, then

$$Y = \lim_{N \to \infty} \frac{1}{N} \sum_{i=0}^{i<N} X_i \qquad (5.39)$$

follows a Gaussian distribution with mean and variance:

$$\mu = \lim_{N \to \infty} \frac{1}{N} \sum_{i=0}^{i<N} \mu_i \qquad (5.40)$$

$$\sigma^2 = \lim_{N \to \infty} \frac{1}{N} \sum_{i=0}^{i<N} \sigma_i^2 \qquad (5.41)$$

We can numerically verify this for the simple case in which X_i are uniform random variables with mean equal to 0:

Listing 5.10: in file: nlib.py

```
1 >>> def added_uniform(n): return sum([random.uniform(-1,1) for i in xrange(n)])/
     n
2 >>> def make_set(n,m=10000): return [added_uniform(n) for j in xrange(m)]
3 >>> Canvas(title='Central Limit Theorem',xlab='y',ylab='p(y)'
4 ...           ).hist(make_set(1),legend='N=1').save('images/central1.png')
5 >>> Canvas(title='Central Limit Theorem',xlab='y',ylab='p(y)'
6 ...           ).hist(make_set(2),legend='N=2').save('images/central3.png')
7 >>> Canvas(title='Central Limit Theorem',xlab='y',ylab='p(y)'
8 ...           ).hist(make_set(4),legend='N=4').save('images/central4.png')
9 >>> Canvas(title='Central Limit Theorem',xlab='y',ylab='p(y)'
10 ...          ).hist(make_set(8),legend='N=8').save('images/central8.png')
```

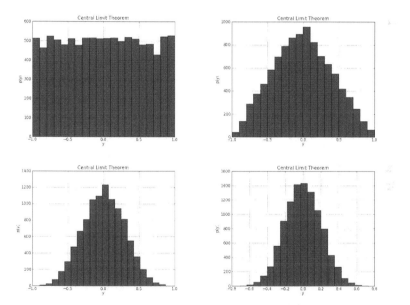

Figure 5.1: Example of distributions for sums of 1, 2, 4, and 8 uniform random variables. The more random variables are added, the better the result approximates a Gaussian distribution.

This theorem is of fundamental importance for stochastic calculus. Notice that the theorem does not apply when the X_i follow distributions that do not have a finite mean or a finite variance.

Distributions that do not follow the central limit are called *Levy distribu-*

tions. They are characterized by fat tails for the form

$$p(x) \underset{x\to\infty}{\sim} \frac{1}{|x|^{1+\alpha}}, \qquad 0 < \alpha < 2 \tag{5.42}$$

An example if the Pareto distribution.

5.1.7 Error in the mean

One consequence of the Central Limit Theorem is a useful formula for evaluating the error in the mean. Let's consider the case of N repeated experiments with outcomes X_i. Let's also assume that each X_i is supposed to be equal to an unknown value μ, but in practice, $X_i = \mu + \varepsilon$, where ε is a random variable with Gaussian distribution centered at zero. One could estimate μ by $\mu = E[X] = \frac{1}{N}\Sigma_i X_i$. In this case, statistical error in the mean is given by

$$\delta\mu = \sqrt{\frac{\sigma^2}{N}} \tag{5.43}$$

where $\sigma^2 = E[(X - \mu)^2] = \frac{1}{N}\Sigma_i(X_i - \mu)^2$.

5.2 Combinatorics and discrete random variables

Often, to compute the probability of discrete random variables, one has to confront the problem of calculating the number of possible finite outcomes of an experiment. Often this problem is solved by combinatorics.

5.2.1 Different plugs in different sockets

If we have n different plugs and m different sockets; in how many ways can we place the plugs in the sockets?

- Case 1, $n \geq m$. All sockets will be filled. We consider the first socket, and we can select any of the n plugs (n combinations). We consider the second socket, and we can select any of the remaining $n - 1$ plugs

($n-1$ combinations), and so on, until we are left with no free sockets and $n-m$ unused plugs; therefore there are

$$n!/(n-m)! = n(n-1)(n-2)...(n-m+1) \qquad (5.44)$$

combinations.

- Case 2, $n \le m$. All plugs have to be used. We consider the first plug, and we can select any of the m sockets (m combinations). We consider the second plug, and we can select any of the remaining $m-1$ sockets ($m-1$ combinations), and so on, until we are left with no spare plugs and $m-n$ free sockets; therefore there are

$$m!/(m-n)! = m(m-1)(m-2)...(m-n+1) \qquad (5.45)$$

combinations. Note that if $m = n$ then case 1 and case 2 agree, as expected.

5.2.2 Equivalent plugs in different sockets

If we have n equivalent plugs and m different sockets, in how many ways can we place the plugs in the sockets?

- Case 1, $n \ge m$. All sockets will be filled. We cannot distinguish one combination from the other because all plugs are the same. There is only one combination.

- Case 2, $n \le m$. All plugs have to be used but not all sockets. There are $m!/(m-n)!$ ways to fill the sockets with different plugs, and there are $n!$ ways to arrange the plugs within the same filled sockets. Therefore there are

$$\binom{m}{n} = \frac{m!}{(m-n)!n!} \qquad (5.46)$$

ways to place n equivalent plugs into m different sockets. Note that if $m = n$

$$\binom{n}{n} = \frac{n!}{(n-n)!n!} = 1 \qquad (5.47)$$

in agreement with case 1.

Here is another example. A club has 20 members and has to elect a president, a vice president, a secretary, and a treasurer. In how many different ways can they select the four officeholders? Think of each office as a socket and each person as a plug; therefore the number combination is $20!/(20-4)! \simeq 1.2 \times 10^5$.

5.2.3 Colored cards

We have 52 cards, 26 black and 26 red. We shuffle the cards and pick three.

- What is the probability that they are all red?

$$\text{Prob}(3red) = \frac{26}{52} \times \frac{25}{51} \times \frac{24}{50} = \frac{2}{17} \qquad (5.48)$$

- What is the probability that they are all black?

$$\text{Prob}(3black) = \text{Prob}(3red) = \frac{2}{17} \qquad (5.49)$$

- What is the probability that they are not all black or all red?

$$\begin{align} \text{Prob}(mixture) &= 1 - \text{Prob}(3red \cup 3black) & (5.50) \\ &= 1 - \text{Prob}(3red) - \text{Prob}(3black) & (5.51) \\ &= 1 - 2\frac{2}{17} & (5.52) \\ &= \frac{13}{17} & (5.53) \end{align}$$

Here is an example of how we can simulate the deck of cards using Python to compute an answer to the last questions:

Listing 5.11: in file: `tests.py`

```
>>> def make_deck(): return [color for i in xrange(26) for color in ('red',
    'black')]
>>> def make_shuffled_deck(): return random.shuffle(make_deck())
>>> def pick_three_cards(): return make_shuffled_deck()[:3]
>>> def simulate_cards(n=1000):
...     counter = 0
```

```
6  ...        for k in xrange(n):
7  ...            c = pick_three_cards()
8  ...            if not (c[0]==c[1] and c[1]==c[2]): counter += 1
9  ...        return float(counter)/n
10 >>> print simulate_cards()
```

5.2.4 Gambler's fallacy

The typical error in computing probabilities is mixing a priori probability with information about past events. This error is called the *gambler's fallacy* [45]. For example, we consider the preceding problem. We see the first two cards, and they are both red. What is the probability that the third one is also red?

- **Wrong answer**: The probability that they are all red is Prob(3*red*) = 2/17; therefore the probability that the third one is also 2/17.

- **Correct answer**: Because we know that the first two cards are red, the third card must belong to a set of (26 black cards + 24 red cards); therefore the probability that it is red is

$$\text{Prob}(red) = \frac{24}{24+26} = \frac{12}{25} \tag{5.54}$$

6

Random Numbers and Distributions

In the previous chapters, we have seen how using the Python `random` module, we can generate uniform random numbers. This module can also generate random numbers following other distributions. The point of this chapter is to understand how random numbers are generated.

6.1   Randomness, determinism, chaos and order

Before we proceed further, there are four important concepts that should be defined because of their implications:

- **Randomness** is the characteristic of a process whose outcome is unpredictable (e.g., at the moment I am writing this sentence, I cannot predict the exact time and date when you will be reading it).

- **Determinism** is the characteristic of a process whose outcome can be predicted from the initial conditions of the system (e.g., if I throw a ball from a known position, at a known velocity and in a known direction, I can predict—calculate—its entire future trajectory).

- **Chaos** is the emergence of randomness from order [46] (e.g., if I am on the top of a hill and I throw the ball in a vertical direction, I cannot predict on which side of the hill it is going to end up). Even if the equations that describe a phenomenon are known and are deterministic, it may happen that a small variation in the initial conditions causes a large difference in the possible deterministic evolution of the system. Therefore the outcome of a process may depend on a tiny variation of the initial parameters. These variations may not be measurable in practice, thus making the process unpredictable and chaotic. Chaos is generally regarded as a characteristic of some differential equations.

- **Order** is the opposite of chaos. It is the emergence of regular and reproducible patterns from a process that, in itself, may be random or chaotic (e.g., if I keep throwing my ball in a vertical direction from the top of a hill and I record the final location of the ball, I eventually find a regular pattern, a probability distribution associated with my experiment, which depends on the direction of the wind, the shape of the hill, my bias in throwing the ball, etc.).

These four concepts are closely related, and they do not necessarily come in opposite pairs as one would expect.

A deterministic process may cause chaos. We can use chaos to generate randomness (we will see examples when covering random number generation). We can study randomness and extract its ordered properties (probability distributions), and we can use randomness to solve deterministic problems (Monte Carlo) such as computing integrals and simulating a system.

6.2 Real randomness

Note that randomness does not necessarily come from chaos. Randomness exists in nature [47][48]. For example, a radioactive atom "decays" into a different atom at some random point in time. For example, an atom of carbon 14 decays into nitrogen 14 by emitting an electron and a neutrino

$$^{14}_{6}C \longrightarrow ^{14}_{7}N + e^- + \bar{\nu}_e \qquad (6.1)$$

at some random time t; t is unpredictable. It can be proven that the randomness in the nuclear decay time is not due to any underlying deterministic process. In fact, constituents of matter are described by quantum physics, and randomness is a fundamental characteristic of quantum systems. Randomness is not a consequence of our ignorance.

This is not usually the case for macroscopic systems. Typically the randomness we observe in some macroscopic systems is not always a consequence of microscopic randomness. Rather, order and determinism emerge from the microscopic randomness, while chaos originates from the complexity of the system.

Because randomness exists in nature, we can use it to produce random numbers with any desired distribution. In particular, we want to use the randomness in the decay time of radioactive atoms to produce random numbers with uniform distribution. We assemble a system consisting of many atoms, and we record the time when we observe atoms decay:

$$t_0, t_1, t_2, t_3, t_4, t_5, \ldots \tag{6.2}$$

One could study the probability distribution of the t_i and find that it follows an exponential probability distribution like

$$\text{Prob}(t_i = t) = \lambda e^{-\lambda t} \tag{6.3}$$

where $t_0 = 1/\lambda$ is the decay time characteristic of the particular type of atom. One characteristic of this distribution is that it is a memoryless process: t_i does not depend on t_{i-1} and therefore the probability that $t_i > t_{i-1}$ is the same as the probability that $t_i < t_{i-1}$.

6.2.1 Memoryless to Bernoulli distribution

Given the sequence $\{t_i\}$ with exponential distribution, we can build a random sequence of zeros and ones (Bernoulli distribution) by applying the following formula, known as the Von Neumann procedure [49]:

$$x_i = \begin{cases} 1 & \text{if } t_i > t_{i-1} \\ 0 & \text{otherwise} \end{cases} \tag{6.4}$$

Note that the procedure can be applied to map any random sequence into a Bernoulli sequence even if the numbers in the original sequence do not follow an exponential distribution, as long as t_i is independent of t_j for any $j < i$ (memoryless distribution).

6.2.2 Bernoulli to uniform distribution

To map a Bernoulli distribution into a uniform distribution, we need to determine the precision (resolution in number of bits) of the numbers we wish to generate. In this example, we will assume 8 bits.

We can think of each number as a point in a $[0,1)$ segment. We generate the uniform number by making a number of choices: we break the segment in two and, according to the value of the binary digit (0 or 1), we select the first part or the second part and repeat the process on the subsegment. Because at each stage we break the segment into two parts of equal length and we select one or the other with the same probability, the final distribution of the selected point is uniform. As an example, we consider the Bernoulli sequence

$$01011110110101010111011010 \qquad (6.5)$$

and we perform the following steps:

- break the sequence into chunks of 8 bits

$$01011110\text{-}11010101\text{-}01110110\text{-}..... \qquad (6.6)$$

- map each chunk $a_0 a_1 a_2 a_3 a_4 a_5 a_6 a_7$ into $x = \sum_{k=0}^{k<8} a_k / 2^{k+1}$ thus obtaining:

$$0.3671875 - 0.83203125 - 0.4609375 - ... \qquad (6.7)$$

A uniform random number generator is usually the first step toward building any other random number generator.

Other physical processes can be used to generate real random numbers using a similar process. Some microprocessors can generate random numbers from random temperature fluctuations. An unpredictable source of randomness is called an *entropy source*.

6.3   Entropy generators

The Linux/Unix operating system provides its own entropy source accessible via "/dev/urandom." This data source is available in Python via `os.urandom()`.

Here we define a class that can access this entropy source and use it to generate uniform random numbers. It follows the same process outlined for the radioactive days:

```
1  class URANDOM(object):
2      def __init__(self, data=None):
3          if data: open('/dev/urandom','wb').write(str(data))
4      def random(self):
5          import os
6          n = 16
7          random_bytes =  os.urandom(n)
8          random_integer = sum(ord(random_bytes[k])*256**k for k in xrange(n))
9          random_float = float(random_integer)/256**n
```

Notice how the constructor allows us to further randomize the data by contributing input to the entropy source. Also notice how the `random()` method reads 16 bites from the stream (using `os.urandom()`), converts each into 8-bit integers, combines them into a 128-bit integer, and then converts it to a float by dividing by 256^{16}.

6.4   Pseudo-randomness

In many cases we do not have a physical device to generate random numbers, and we require a software solution. Software is deterministic, the outcome is reproducible, therefore it cannot be used to generate randomness, but it can generate pseudo-randomness. The outputs of pseudo random number generators are not random, yet they may be considered random for practical purposes. John von Neumann observed in 1951 that "anyone who considers arithmetical methods of producing random digits is, of course, in a state of sin." (For attempts to generate "truly random" numbers, see the article on hardware random number generators.) Nevertheless, pseudo random numbers are a critical part of modern computing, from cryptography to the Monte Carlo method for simulating physical systems.

Pseudo random numbers are relatively easy to generate with software, and they provide a practical alternative to random numbers. For some applications, this is adequate.

6.4.1 Linear congruential generator

Here is probably the simplest possible pseudo random number generator:

$$x_i = (ax_{i-1} + c)\mod m \quad (6.8)$$
$$y_i = x_i/m \quad (6.9)$$

With the choice $a = 65539$, $c = 0$, and $m = 2^{31}$, this generator is called RANDU. It is of historical importance because it is implemented in the C rand() function. The RANDU generator is particularly fast because the modulus can be implemented using the finite 32-bit precision.

Here is a possible implementation for $c = 0$:

Listing 6.1: in file: nlib.py

```
class MCG(object):
    def __init__(self,seed,a=65539,m=2**31):
        self.x = seed
        self.a, self.m = a, m
    def next(self):
        self.x = (self.a*self.x) % self.m
        return self.x
    def random(self):
        return float(self.next())/self.m
```

which we can test with

```
>>> randu = MCG(seed=1071914055)
>>> for i in xrange(10): print randu.random()
...
```

The output numbers "look" random but are not truly random. Running the same code with the same seed generates the same output. Notice the following:

- PRNGs are typically implemented as a recursive expression that, given x_{i-1}, produces x_i.

- PRNGs have to start from an initial value, x_0, called the *seed*. A typical choice is to set the seed equal to the number of seconds from the conventional date and time "Thu Jan 01 01:00:00 1970." This is not always a good choice.

- PRNGs are periodic. They generate numbers in a finite set and then they repeat themselves. It is desirable to have this set as large as possible.

- PRNGs depend on some parameters (e.g., a and m). Some parameter choices lead to trivial random number generators. In general, some choices are better than others, and a few are optimal. In particular, the values of a and m determine the period of the random number generator. An optimal choice is the one with the longest period.

For a linear congruential generator, because of the mod operation, the period is always less than or equal to m. When c is nonzero, the period is equal to m only if c and m are relatively prime, $a - 1$ is divisible by all prime factors of m, and $a - 1$ is a multiple of 4 when m is a multiple of 4.

In the case of RANDU, the period is $m/4$. A better choice is using $a = 7^5$ and $m = 2^{31} - 1$ (known as the Marsenne prime number) because it can be proven that m is in fact the period of the generator:

$$x_i = (7^5 x_{i-1}) \mathrm{mod}(2^{31} - 1) \tag{6.10}$$

Here are some examples of MCG used by various systems:

Source	m	a	c
Numerical Recipes	2^{32}	1664525	1013904223
glibc (used by GCC)	2^{32}	1103515245	12345
Apple CarbonLib	$2^{31} - 1$	16807	0
java.util.Random	2^{48}	25214903917	11

When c is set to zero, a linear congruential generator is also called a multiplicative congruential generator.

6.4.2 Defects of PRNGs

The non-randomness of pseudo random number generators manifests itself in at least two different ways:

- The sequence of generated numbers is periodic, therefore only a finite set of numbers can come out of the generator, and many of the numbers will never be generated. This is not a major problem if the period is much larger (some order of magnitude) than the number of random numbers needed in the Monte Carlo computation.

- The sequence of generated numbers presents bias in the form of "patterns." Sometimes these patterns are evident, sometimes they are not evident. Patterns exist because the pseudo random numbers are not random but are generated using a recursive formula. The existence of these patterns may introduce a bias in Monte Carlo computations that use the generator. This is a nasty problem, and the implications depend on the specific case.

An example of pattern/bias is discussed in ref. [51] and can be seen in fig. 6.4.2.

6.4.3 Multiplicative recursive generator

Another modification of the multiplicative congruential generator is the following:

$$x_i = (a_1 x_{i-1} + a_2 x_{i-2} + ... + a_k x_{i-k}) \bmod m \qquad (6.11)$$

The advantage of this generator is that if m is prime, the period of this type of generator can be as big as $m^k - 1$. This is much larger than a simple multiplicative congruential generator.

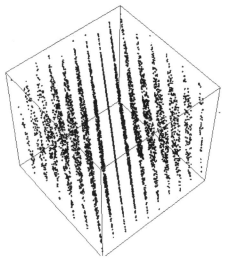

Figure 6.1: In this plot, each three consecutive random numbers (from RANDU) are interpreted as (x, y, z) coordinates of a random point. The image clearly shows the points are not distributed at random. Image from ref. [51].

An example is $a_1 = 107374182$, $a_2 = a_3 = a_4 = 0$, $a_5 = 104480$, and $m = 2^{31} - 1$, where the period is

$$(2^{31} - 1)^5 - 1 \simeq 4.56 \times 10^{46} \tag{6.12}$$

6.4.4   Lagged Fibonacci generator

$$x_i = (x_{i-j} + x_{i-k}) \bmod m \tag{6.13}$$

This is similar to the multiplicative recursive generator earlier. If m is prime and $j \neq k$, the period can be as large as $m^k - 1$.

6.4.5   Marsaglia's add-with-carry generator

$$x_i = (x_{i-j} + x_{i-k} + c_i) \bmod m \tag{6.14}$$

where $c_1 = 0$ and $c_i = 1$ if $(x_{i-1-j} + x_{i-1-k} + c_{i-1}) < m$, 0

254 ANNOTATED ALGORITHMS IN PYTHON

6.4.6 Marsaglia's subtract-and-borrow generator

$$x_i = (x_{i-j} - x_{i-k} - c_i) \bmod m \qquad (6.15)$$

where $k > j > 0$, $c_1 = 0$, and $c_i = 1$ if $(x_{i-1-j} - x_{i-1-k} - c_{i-1}) < 0$, 0 otherwise.

6.4.7 Lüscher's generator

The Marsaglia's subtract-and-borrow is a very popular generator, but it is known to have some problems. For example, if we construct vector

$$v_i = (x_i, x_{i+1}, ..., x_{i+k}) \qquad (6.16)$$

and the coordinates of the point v_i are numbers closer to each other then the coordinates of the point v_{i+k} are also close to each other. This indicates that there is an unwanted correlation between the points $x_i, x_{i+1}, ..., x_{i+k}$. Lüscher observed [50] that the Marsaglia's subtract-and-borrow is equivalent to a chaotic discrete dynamic system, and the preceding correlation dies off for points that distance themselves more than k. Therefore he proposed to modify the generator as follows: instead of taking all x_i numbers, read k successive elements of the sequence, discard $p - k$ numbers, read k numbers, and so on. The number p has to be chosen to be larger than k. When $p = k$, the original Marsaglia generator is recovered.

6.4.8 Knuth's polynomial congruential generator

$$x_i = (ax_{i-1}^2 + bx_{i-1} + c) \bmod m \qquad (6.17)$$

This generator takes the form of a more complex function. It makes it harder to guess one number in the sequence from the following numbers; therefore it finds applications in cryptography.

Another example is the Blum, Blum, and Shub generator:

$$x_i = x_{i-1}^2 \bmod m \qquad (6.18)$$

6.4.9 PRNGs in cryptography

Random numbers find many applications in cryptography. For example, consider the problem of generating a random password, a digital signature, or random encryption keys for the Diffie–Hellmann and the RSA encryption schemes.

A cryptographically secure pseudo random number generator (CSPRNG) is a pseudo random number generator (PRNG) with properties that make it suitable for use in cryptography.

In addition to the normal requirements for a PRNG (that its output should pass all statistical tests for randomness) a CSPRNG must have two additional properties:

- It should be difficult to predict the output of the CSPRNG, wholly or partially, from examining previous outputs.

- It should be difficult to extract all or part of the internal state of the CSPRNG from examining its output.

Most PRNGs are not suitable for use as CSPRNGs. They must appear random in statistical tests, but they are not designed to resist determined mathematical reverse engineering.

CSPRNGs are designed explicitly to resist reverse engineering. There are a number of examples of CSPRNGs. Blum, Blum, and Shub has the strongest security proofs, though it is slow.

Many pseudo random number generators have the form

$$x_i = f(x_{i-1}, x_{i-2}, ..., x_{i-k}) \qquad (6.19)$$

For example, the next random number depends on the past k numbers. Requirements for CSPRNGs used in cryptography are that

- Given $x_{i-1}, x_{i-2}, ..., x_{i-k}$, x_i can be computed in polynomial time, while

- Given $x_i, x_{i-2}, ..., x_{i-k}$, x_{i-1} must not be computable in polynomial time.

The first requirement means that the PRNG must be fast. The second requirement means that if a malicious agent discovers a random number used as a key, he or she cannot easily compute all previous keys generated using the same PRNG.

6.4.10 Inverse congruential generator

$$x_i = (ax_{i-1}^{-1} + c) \bmod m \tag{6.20}$$

where x_{i-1}^{-1} is the multiplicative inverse of x_{i-1} modulo m, for example, $x_{i-1}x_{i-1}^{-1} = 1 \bmod m$.

6.4.11 Marsenne twister

One of the best PRNG algorithms (because of its long period, uniform distribution, and speed) is the Marsenne twister, which produces a 53-bit random number, and it has a period of $2^{19937} - 1$ (this number is 6002 digits long!). The Python random module uses the Marsenne twister. Although discussing the inner working of this algorithm is beyond the scope of these notes, we provide a pure Python implementation of the Marsenne twister:

Listing 6.2: in file: nlib.py

```python
class MarsenneTwister(object):
    """
    based on:
    Knuth 1981, The Art of Computer Programming
    Vol. 2 (2nd Ed.), pp102]
    """
    def __init__(self,seed=4357):
        self.w = []    # the array for the state vector
        self.w.append(seed & 0xffffffff)
        for i in xrange(1, 625):
            self.w.append((69069 * self.w[i-1]) & 0xffffffff)
        self.wi = i
    def random(self):
        w = self.w
        wi = self.wi
        N, M, U, L = 624, 397, 0x80000000, 0x7fffffff
        K = [0x0, 0x9908b0df]
        y = 0
        if wi >= N:
```

```
20      for kk in xrange((N-M) + 1):
21          y = (w[kk]&U)|(w[kk+1]&L)
22          w[kk] = w[kk+M] ^ (y >> 1) ^ K[y & 0x1]
23
24      for kk in xrange(kk, N):
25          y = (w[kk]&U)|(w[kk+1]&L)
26          w[kk] = w[kk+(M-N)] ^ (y >> 1) ^ K[y & 0x1]
27      y = (w[N-1]&U)|(w[0]&L)
28      w[N-1] = w[M-1] ^ (y >> 1) ^ K[y & 0x1]
29  wi = 0
30  y = w[wi]
31  wi += 1
32  y ^= (y >> 11)
33  y ^= (y << 7) & 0x9d2c5680
34  y ^= (y << 15) & 0xefc60000
35  y ^= (y >> 18)
36  return (float(y)/0xffffffff )
```

In the above code, numbers starting with 0x are represented in hexadec-imal notation. The symbols &, ^, <<, and >> are bitwise operators. & is a binary AND, ^ is a binary exclusive XOR, << shifts all bits to the left and >> shifts all bits to the right. We refer to the Python official documentation for details.

6.5   Parallel generators and independent sequences

It is often necessary to generate many independent sequences.

For example, you may want to generate streams or random numbers in parallel using multiple machines and processes, and you need to ensure that the streams do not overlap.

A common mistake is to generate the sequences using the same generator with different seeds. This is not a safe procedure because it is not obvious if the seed used to generate one sequence belongs to the sequence gener-ated by the other seed. The two sequences of random numbers are not independent, but they are merely shifted in respect to each other.

For example, here are two RANDU sequences generated with different but dependent seeds:

seed	1071931562	50554362	
y_0	0.252659081481	0.867315522395	
y_1	0.0235412092879	0.992022250779	
y_2	0.867315522395	0.146293803118	(6.21)
y_3	0.992022250779	0.949562561698	
y_4	0.146293803118	0.380731142126	
y_5	...	...	

Note that the second sequence is the same as the first but shifted by two lines.

Three standard techniques for generating independent sequences are non-overlapping blocks, leapfrogging, and Lehmer trees.

6.5.1   Non-overlapping blocks

Let's consider one sequence of pseudo random numbers:

$$x_0, x_1, ..., x_k, x_{k+1}, ..., x_{2k}, x_{2k+1}, ..., x_{3k}, x_{3k+1}, ..., \qquad (6.22)$$

One can break it into subsequences of k numbers:

$$x_0, x_1, ..., x_{k-1} \qquad (6.23)$$

$$x_k, x_{k+1}, ..., x_{2k-1} \qquad (6.24)$$

$$x_{2k}, x_{2k+1}, ..., x_{3k-1} \qquad (6.25)$$

$$... \qquad (6.26)$$

If the original sequence is created with a multiplicative congruential generator

$$x_i = a x_{i-1} \bmod m \qquad (6.27)$$

the subsequences can be generated independently because

$$x_{nk-1} = a^{nk-1} x_0 \bmod m \qquad (6.28)$$

if the seed of the arbitrary sequence is $x_{nk}, x_{nk+1}, ..., x_{nk-1}$. This is particularly convenient for parallel computers where one computer generates the seeds for the subsequences and the processing nodes, independently, generated the subsequences.

6.5.2 Leapfrogging

Another and probably more popular technique is leapfrogging. Let's consider one sequence of pseudo random numbers:

$$x_0, x_1, ..., x_k, x_{k+1}, ..., x_{2k}, x_{2k+1}, ..., x_{3k}, x_{3k+1}, ..., \tag{6.29}$$

One can break it into subsequences of k numbers:

$$x_0, x_k, x_{2k}, x_{3k}, ... \tag{6.30}$$

$$x_1, x_{1+k}, x_{1+2k}, x_{1+3k}, ... \tag{6.31}$$

$$x_2, x_{2+k}, x_{2+2k}, x_{2+3k}, ... \tag{6.32}$$

$$... \tag{6.33}$$

The seeds $x_1, x_2, ..x_{k-1}$ are generated from x_0, and the independent sequences can be generated independently using the formula

$$x_{i+k} = a^k x_i \bmod m \tag{6.34}$$

Therefore leapfrogging is a viable technique for parallel random number generators.

Here is an example of a usage of `leapfrog`:

Listing 6.3: in file: `nlib.py`

```
def leapfrog(mcg,k):
    a = mcg.a**k % mcg.m
    return [MCG(mcg.next(),a,mcg.m) for i in range(k)]
```

Here is an example of usage:

```
>>> generators=leapfrog(MCG(m),3)
>>> for k in xrange(3):
...     for i in xrange(5):
...         x=generators[k].random()
...         print k,'\t',i,'\t',x
```

The Marsenne twister algorithm implemented in `os.random` has leapfrogging built in. In fact, the module includes a `random.jumpahead(n)` method that allows us to efficiently skip n numbers.

6.5.3   Lehmer trees

Lehmer trees are binary trees, generated recursively, where each node contains a random number. We start from the root containing the seed, x_0, and we append two children containing, respectively,

$$x_i^L = (a_L x_{i-1} + c_L) \bmod m \qquad (6.35)$$
$$x_i^R = (a_R x_{i-1} + c_R) \bmod m \qquad (6.36)$$

then, recursively, append nodes to the children.

6.6   Generating random numbers from a given distribution

In this section and the next, we provide examples of distributions other than uniform and algorithms to generate numbers using these distributions. The general strategy consists of finding ways to map uniform random numbers into numbers following a different distribution. There are two general techniques for mapping uniform into nonuniform random numbers:

- accept–reject (applies to both discrete and continuous distributions)

- inversion methods (applies to continuous distributions only)

Consider the problem of generating a random number x from a given distribution $p(x)$. The *accept–reject method* consists of generating x using a different distribution, $g(x)$, and a uniform random number, u, between 0,1. If $u < p(x)/Mg(x)$ (M is the max of $p(x)/g(x)$), then x is the desired random number following distribution $p(x)$. If not, try another number.

To visualize why this works, imagine graphing the distribution p of the random variable x onto a large rectangular board and throwing darts at it, the coordinates of the dart being (x, u). Assume that the darts are uniformly distributed around the board. Now take off (reject) all of the darts that are outside the curve. The remaining darts will be distributed uniformly within the curve, and the x-positions of these darts will be distributed according to the random variable's density. This is because there is the most room for the darts to land where the curve is highest and thus the probability density is greatest.

The g distribution is nothing but a shape so that all darts we throw are below it. There are two particular cases. In one case, $g = p$, we only throw darts below the p that we want; therefore we accept them all. This is the most efficient case, but it is not of practical interest because it means the accept–reject is not doing anything, as we already know now to generate numbers according to p. The other case is $g(x) = constant$. This means we generate the x uniformly before the accept–reject. This is equivalent to throwing the darts everywhere on the square board, without even trying to be below the curve p.

The inversion method instead is more efficient but requires some math. It states that if $F(x)$ is a cumulative distribution function and u is a uniform random number between 0 and 1, then $x = F^{-1}(u)$ is a random number with distribution $p(x) = F'(x)$. For those distributions where F can be expressed in analytical terms and inverted, the inversion method is the best way to generate random numbers. An example is the exponential distribution.

We will create a new class RandomSource that includes methods to generate the random number.

6.6.1   Uniform distribution

The uniform distributions are simple probability distributions which, in the discrete case, can be characterized by saying that all possible values are equally probable. In the continuous case, one says that all intervals of the same length are equally probable.

There are two types of uniform distribution: discrete and continuous.

Here we consider the discrete case as we implement it into a randint method:

Listing 6.4: in file: nlib.py

```
class RandomSource(object):
```

```
2      def __init__(self,generator=None):
3          if not generator:
4              import random as generator
5          self.generator = generator
6      def random(self):
7          return self.generator.random()
8      def randint(self,a,b):
9          return int(a+(b-a+1)*self.random())
```

Notice that the random `RandomSource` constructor expects a generator such as `MCG`, `MarsenneTwister`, or simply `random` (default value). The `random()` method is a proxy method for the equivalent method of the underlying generator object.

We can use `randint` to generate a random choice from a finite set when each option has the same probability:

Listing 6.5: in file: `nlib.py`

```
1      def choice(self,S):
2          return S[self.randint(0,len(S)-1)]
```

6.6.2 Bernoulli distribution

The Bernoulli distribution, named after Swiss scientist James Bernoulli, is a discrete probability distribution that takes value 1 with probability of success p and value 0 with probability of failure $q = 1 - p$:

$$p(k) \equiv \left\{ \begin{array}{ll} p & \text{if } k = 1 \\ 1 - p & \text{if } k = 0 \\ 0 & \text{if not } k \in \{0, 1\} \end{array} \right\} \qquad (6.37)$$

A Bernoulli random variable has an expected value of p and variance of pq.

We implement it by adding a corresponding method to the `RandomSource` class:

Listing 6.6: in file: `nlib.py`

```
1      def bernoulli(self,p):
2          return 1 if self.random()<p else 0
```

6.6.3   Biased dice and table lookup

A generalization of the Bernoulli distribution is a distribution in which we have a finite set of choices, each with an associated probability. The table can be a list of tuples (value, probability) or a dictionary of value:probability:

Listing 6.7: in file: nlib.py

```
def lookup(self,table, epsilon=1e-6):
    if isinstance(table,dict): table = table.items()
    u = self.random()
    for key,p in table:
        if u<p+epsilon:
            return key
        u = u - p
    raise ArithmeticError('invalid probability')
```

Let's say we want a random number generator that can only produce the outcome 0,1 or 2 with known probabilities:

$$\text{Prob}(X = 0) = 0.50 \qquad (6.38)$$
$$\text{Prob}(X = 1) = 0.23 \qquad (6.39)$$
$$\text{Prob}(X = 2) = 0.27 \qquad (6.40)$$

Because the probability of the possible outcomes are rational numbers (fractions), we can proceed as follows:

```
>>> def test_lookup(nevents=100,table=[(0,0.50),(1,0.23),(2,0.27)]):
...     g = RandomSource()
...     f=[0,0,0]
...     for k in xrange(nevents):
...         p=g.lookup(table)
...         print p,
...         f[p]=f[p]+1
...     print
...     for i in xrange(len(table)):
...         f[i]=float(f[i])/nevents
...         print 'frequency[%i]=%f' % (i,f[i])
```

which produces the following output:

```
0 1 2 0 0 0 2 2 2 2 0 0 0 2 1 1 2 0 0 2 1 2 0 1
0 0 0 0 0 0 0 0 0 1 2 2 0 0 1 2 2 0 0 1 0 0 1 0
0 0 0 0 0 2 2 0 2 0 2 0 0 0 0 2 1 2 0 2 0 2 0 0 0
0 0 0 2 2 0 0 0 0 2 1 1 0 2 0 0 0 0 0 1 0 1 0 0 0
```

```
5 frequency[0]=0.600000
6 frequency[1]=0.140000
7 frequency[2]=0.260000
```

Eventually, by repeating the experiment many more times, the frequencies of 0,1 and 2 will approach the input probabilities.

Given the output frequencies, what is the probability that they are compatible with the input frequency? The answer to this question is given by the χ^2 and its distribution. We discuss this later in the chapter.

In some sense, we can think of the table lookup as an application of the linear search. We start with a segment of length 1, and we break it into smaller contiguous intervals of length $\text{Prob}(X = 0), \text{Prob}(X = 1), ..., \text{Prob}(x = n - 1)$ so that $\sum \text{Prob}(X = i) = 1$. We then generate a random point on the initial segment, and we ask in which of the n intervals it falls. The table lookup method linearly searches the interval.

This technique is $\Theta(n)$, where n is the number of outcomes of the computation. Therefore it becomes impractical if the number of cases is large. In this case, we adopt one of the two possible techniques: the Fishman–Yarberry method or the accept–reject method.

6.6.4 Fishman–Yarberry method

The Fishman–Yarberry [52] (F-Y) method is an improvement over the naive table lookup that runs in $O(\lceil \log_2 n \rceil)$. As the naive table lookup is an application of the linear search, the F-Y is an application of the binary search.

Let's assume that $n = 2^t$ is an exact power of 2. If this is not the case, we can always reduce to this case by adding new values to the lookup table corresponding to 0 probability. The basic data structure behind the F-Y method is an array of arrays a_{ij} built according to the following rules:

- $\forall j \geq 0, a_{0j} = \text{Prob}(X = x_j)$

- $\forall j \geq 0$ and $i > 0$, $a_{ij} = a_{i-1,2j} + a_{i-1,2j+1}$

Note that $0 \leq i < t$ and $\forall i \geq 0, 0 \leq j < 2^{t-i}$, where $t = \log_2 n$. The array of arrays a can be represented as follows:

$$a_{ij} = \begin{pmatrix} a_{00} & a_{01} & a_{02} & \cdots & a_{0,n-1} \\ \cdots & \cdots & \cdots & \cdots & \\ a_{t-2,0} & a_{t-2,1} & a_{t-2,2} & a_{t-2,3} & \\ a_{t-1,0} & a_{t-1,1} & & & \end{pmatrix} \quad (6.41)$$

In other words, we can say that

- a_{ij} represents the probability

$$\text{Prob}(X = x_j) \quad (6.42)$$

- a_{1j} represents the probability

$$\text{Prob}(X = x_{2j} or X = x_{2j+1}) \quad (6.43)$$

- a_{4j} represents the probability

$$\text{Prob}(X = x_{4j} \text{ or } X = x_{4j+1} \text{ or } X = x_{4j+2} \text{ or } X = x_{4j+3}) \quad (6.44)$$

- a_{ij} represents the probability

$$\text{Prob}(X \in \{x_k | 2^i j \leq k < 2^i(j+1)\}) \quad (6.45)$$

This algorithm works like the binary search, and at each step, it confronts the uniform random number u with a_{ij} and decides if u falls in the range $\{x_k | 2^i j \leq k < 2^i(j+1)\}$ or in the complementary range $\{x_k | 2^i(j+1) \leq k < 2^i(j+2)\}$ and decreases i.

Here is the algorithm implemented as a class member function. The constructor of the class creates an array a once and for all. The method discrete_map maps a uniform random number u into the desired discrete integer:

```
class FishmanYarberry(object):
    def __init__(self,table=[[0,0.2], [1,0.5], [2,0.3]]):
        t=log(len(table),2)
        while t!=int(t):
            table.append([0,0.0])
```

```
 6          t=log(len(table),2)
 7      t=int(t)
 8      a=[]
 9      for i in xrange(t):
10          a.append([])
11          if i==0:
12              for j in xrange(2**t):
13                  a[i].append(table[j,1])
14          else:
15              for j in xrange(2**(t-i)):
16                  a[i].append(a[i-1][2*j]+a[i-1][2*j+1])
17      self.table=table
18      self.t=t
19      self.a=a
20
21  def discrete_map(self, u):
22      i=int(self.t)-1
23      j=0
24      b=0
25      while i>0:
26          if u>b+self.a[i][j]:
27              b=b+self.a[i][j]
28              j=2*j+2
29          else:
30              j=2*j
31          i=i-1
32      if u>b+self.a[i][j]:
33          j=j+1
34      return self.table[j][0]
```

6.6.5 Binomial distribution

The binomial distribution is a discrete probability distribution that describes the number of successes in a sequence of n independent experiments, each of which yields success with probability p. Such a success–failure experiment is also called a Bernoulli experiment.

A typical example is the following: 7% of the population are left-handed. You pick 500 people randomly. How likely is it that you get 30 or more left-handed? The number of left-handed you pick is a random variable X that follows a binomial distribution with $n = 500$ and $p = 0.07$. We are interested in the probability $\text{Prob}(X = 30)$.

In general, if the random variable X follows the binomial distribution with parameters n and p, the probability of getting exactly k successes is given by

$$p(k) = \text{Prob}(X = k) \equiv \binom{n}{k} p^k (1 - p)^{n-k} \tag{6.46}$$

for $k = 0, 1, 2, ..., n$.

The formula can be understood as follows: we want k successes (p^k) and $n - k$ failures ($(1 - p)^{n-k}$). However, the k successes can occur anywhere among the n trials, and there are $\binom{n}{k}$ different ways of distributing k successes in a sequence of n trials.

The mean is $\mu_X = np$, and the variance is $\sigma_X^2 = np(1 - p)$.

If X and Y are independent binomial variables, then $X + Y$ is again a binomial variable; its distribution is

$$p(k) = \text{Prob}(X = k) = \binom{n_X + n_Y}{k} p^k (1 - p)^{n-k} \tag{6.47}$$

We can generate random numbers following binomial distribution using a table lookup with table

$$\text{table}[k] = \text{Prob}(X = k) = \binom{n}{k} p^k (1 - p)^{n-k} \tag{6.48}$$

For large n, it may be convenient to avoid storing the table and use the formula directly to compute its elements on a need-to-know basis. Moreover, because the table is accessed sequentially by the table lookup algorithm, one may just notice that the current recursive relation holds:

$$\text{Prob}(X = 0) = (1 - p)^n \tag{6.49}$$

$$\text{Prob}(X = k + 1) = \frac{n}{k+1} \frac{p}{1 - p} \text{Prob}(X = k) \tag{6.50}$$

This allows for a very efficient implementation:

Listing 6.8: in file: `nlib.py`

```
def binomial(self,n,p,epsilon=1e-6):
```

```
2     u = self.random()
3     q = (1.0-p)**n
4     for k in xrange(n+1):
5         if u<q+epsilon:
6             return k
7         u = u - q
8         q = q*(n-k)/(k+1)*p/(1.0-p)
9     raise ArithmeticError('invalid probability')
```

6.6.6 Negative binomial distribution

In probability theory, the negative binomial distribution is the probability distribution of the number of trials n needed to get a fixed (nonrandom) number of successes k in a Bernoulli process. If the random variable X is the number of trials needed to get r successes in a series of trials where each trial has probability of success p, then X follows the negative binomial distribution with parameters r and p:

$$p(n) = \text{Prob}(X = n) = \binom{n-1}{k-1} p^k (1 - p)^{n-k} \qquad (6.51)$$

Here is an example:

John, a kid, is required to sell candy bars in his neighborhood to raise money for a field trip. There are thirty homes in his neighborhood, and he is told not to return home until he has sold five candy bars. So the boy goes door to door, selling candy bars. At each home he visits, he has a 0.4 probability of selling one candy bar and a 0.6 probability of selling nothing.

- What's the probability of selling the last candy bar at the nth house?

$$p(n) = \binom{n-1}{4} 0.4^5 0.6^{n-5} \qquad (6.52)$$

- What's the probability that he finishes on the tenth house?

$$p(10) = \binom{9}{4} 0.4^5 0.6^5 = 0.10 \qquad (6.53)$$

- What's the probability that he finishes on or before reaching the eighth house? Answer: To finish on or before the eighth house, he must finish at the fifth, sixth, seventh, or eighth house. Sum those probabilities:

$$\sum_{i=5,6,7,8} p(i) = 0.1737 \qquad (6.54)$$

- What's the probability that he exhausts all houses in the neighborhood without selling the five candy bars?

$$1 - \sum_{i=5,\dots,30} p(i) = 0.0015 \qquad (6.55)$$

As we the binomial distribution, we can find an efficient recursive formula for the negative binomial distribution:

$$\text{Prob}(X = k) = p^k \qquad (6.56)$$

$$\text{Prob}(X = n+1) = \frac{n}{n-k+1}(1-p)\text{Prob}(X=n) \qquad (6.57)$$

This allows for a very efficient implementation:

Listing 6.9: in file: `nlib.py`

```
def negative_binomial(self,k,p,epsilon=1e-6):
    u = self.random()
    n = k
    q = p**k
    while True:
        if u<q+epsilon:
            return n
        u = u - q
        q = q*n/(n-k+1)*(1-p)
        n = n + 1
    raise ArithmeticError('invalid probability')
```

Notice once again that, unlike the binomial case, here k is fixed, not n, and the random variable has a minimum value of k but no upper bound.

6.6.7 Poisson distribution

The Poisson distribution is a discrete probability distribution discovered by Siméon-Denis Poisson. It describes a random variable X that counts, among other things, the number of discrete occurrences (sometimes called *arrivals*) that take place during a time interval of given length. The probability that there are exactly x occurrences (x being a natural number including 0, $k = 0, 1, 2, ...$) is

$$p(k) = \text{Prob}(X = k) = e^{-\lambda}\frac{\lambda^k}{k!} \qquad (6.58)$$

The Poisson distribution arises in connection with Poisson processes. It applies to various phenomena of discrete nature (i.e., those that may happen 0, 1, 2, 3,...., times during a given period of time or in a given area) whenever the probability of the phenomenon happening is constant in time or space. The Poisson distribution differs from the other distributions considered in this chapter because it is different than zero for any natural number k rather than for a finite set of k values.

Examples include the following:

- The number of unstable nuclei that decayed within a given period of time in a piece of radioactive substance.

- The number of cars that pass through a certain point on a road during a given period of time.

- The number of spelling mistakes a secretary makes while typing a single page.

- The number of phone calls you get per day.

- The number of times your web server is accessed per minute.

- The number of roadkill you find per mile of road.

- The number of mutations in a given stretch of DNA after a certain amount of radiation.

- The number of pine trees per square mile of mixed forest.

• The number of stars in a given volume of space.

The limit of the binomial distribution with parameters n and $p = \lambda/n$, for n approaching infinity, is the Poisson distribution:

$$\frac{n!}{(n-k)!k!}\left(\frac{\lambda}{n}\right)^k\left(1-\frac{\lambda}{n}\right)^{n-k} \simeq e^{-\lambda}\frac{\lambda^k}{k!} + O(\frac{1}{n}) \qquad (6.59)$$

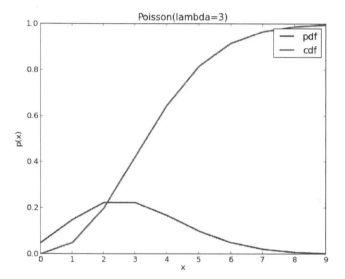

Figure 6.2: Example of Poisson distribution.

Intuitively, the meaning of λ is the following:

Let's consider a unitary time interval T and divide it into n subintervals of the same size. Let p_n be the probability of one success occurring in a single subinterval. For T fixed when $n \to \infty$, $p_n \to 0$ but the limit

$$\lim_{n\to\infty} p_n \qquad (6.60)$$

is finite. This limit is λ.

We can use the same technique adopted for the binomial distribution and observe that for Poisson,

$$\text{Prob}(X \ = \ 0) = e^{-\lambda} \tag{6.61}$$

$$\text{Prob}(X \ = \ k+1) = \frac{\lambda}{k+1}\text{Prob}(X = k) \tag{6.62}$$

therefore the preceding algorithm can be modified into

Listing 6.10: in file: `nlib.py`

```
def poisson(self,lamb,epsilon=1e-6):
    u = self.random()
    q = exp(-lamb)
    k=0
    while True:
        if u<q+epsilon:
            return k
        u = u - q
        q = q*lamb/(k+1)
        k = k+1
    raise ArithmeticError('invalid probability')
```

Note how this algorithm may take an arbitrary amount of time to generate a Poisson distributed random number, but eventually it stops. If u is very close to 1, it is possible that errors due to finite machine precision cause the algorithm to enter into an infinite loop. The $+\varepsilon$ term can be used to correct this unwanted behavior by choosing ε relatively small compared with the precision required in the computation, but larger than machine precision.

6.7 Probability distributions for continuous random variables

6.7.1 Uniform in range

A typical problem is generating random integers in a given range $[a, b]$, including the extreme. We can map uniform random numbers $y_i \in (0, 1)$ into integers by using the formula

$$h_i = a + \lfloor (b - a + 1)y_i \rfloor \tag{6.63}$$

Listing 6.11: in file: `nlib.py`

```
def uniform(self,a,b):
    return a+(b-a)*self.random()
```

6.7.2 Exponential distribution

The exponential distribution is used to model Poisson processes, which are situations in which an object initially in state A can change to state B with constant probability per unit time λ. The time at which the state actually changes is described by an exponential random variable with parameter λ. Therefore the integral from 0 to T over $p(t)$ is the probability that the object is in state B at time T.

The probability mass function is given by

$$p(x) = \lambda e^{-\lambda x} \qquad (6.64)$$

The exponential distribution may be viewed as a continuous counterpart of the geometric distribution, which describes the number of Bernoulli trials necessary for a discrete process to change state. In contrast, the exponential distribution describes the time for a continuous process to change state.

Examples of variables that are approximately exponentially distributed are as follows:

- the time until you have your next car accident
- the time until you get your next phone call
- the distance between mutations on a DNA strand
- the distance between roadkill

An important property of the exponential distribution is that it is memoryless: the chance that an event will occur s seconds from now does not depend on the past. In particular, it does not depend on how much time we have been waiting already. In a formula we can write this condition as

$$\text{Prob}(X > s + t | X > t) = \text{Prob}(X > s) \qquad (6.65)$$

Any process satisfying the preceding condition is a Poisson process. The number of events per time unit is given by the Poisson distribution, and the time interval between consecutive events is described by the exponential distribution.

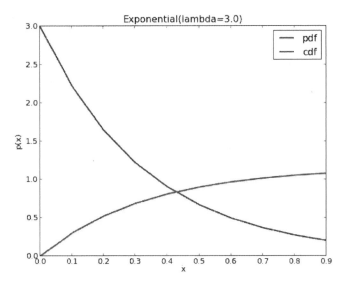

Figure 6.3: Example of exponential distribution.

The exponential distribution can be generated using the inversion method. The scope is to determine a function $x = f(u)$ that maps a uniformly distributed variable u into a continuous random variable x with probability mass function $p(x) = \lambda e^{-\lambda x}$.

According to the inversion method, we proceed by computing F:

$$F(x) = \int_0^x p(y)dy = 1 - e^{-\lambda x} \qquad (6.66)$$

and we then invert $u = F(x)$, thus obtaining

$$x = -\frac{1}{\lambda}\log(1-u) \tag{6.67}$$

Now notice that if u is uniform, $1 - u$ is also uniform; therefore we can further simplify:

$$x = -\frac{1}{\lambda}\log u \tag{6.68}$$

We implement as follows:

Listing 6.12: in file: nlib.py

```
def exponential(self,lamb):
    return -log(self.random())/lamb
```

This is an important distribution, and Python has a function for it:

```
random.expovariate(lamb)
```

6.7.3   Normal/Gaussian distribution

The normal distribution (also known as *Gaussian distribution*) is an extremely important probability distribution considered in statistics. Here is the probability mass function:

$$p(x) = \frac{1}{\sigma\sqrt{2\pi}}e^{-\frac{(x-\mu)^2}{2\sigma^2}} \tag{6.69}$$

where $E[X] = \mu$ and $E[(x-\mu)^2] = \sigma^2$.

The standard normal distribution is the normal distribution with a mean of 0 and a standard deviation of 1. Because the graph of its probability density resembles a bell, it is often called the *bell curve*.

The Gaussian distribution has two important properties:

- The average of many independent random variables with finite mean and finite variance tends to be a Gaussian distribution.

- The sum of two independent Gaussian random variables with means μ_1 and μ_2 and variances σ_1^2 and σ_2^2 is also a Gaussian random variable with mean $\mu = \mu_1 + \mu_2$ and variance $\sigma^2 = \sigma_1^2 + \sigma_2^2$.

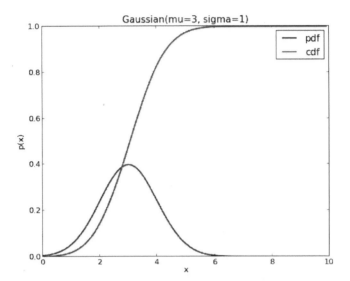

Figure 6.4: Example of Gaussian distribution.

There is no way to map a uniform random number into a Gaussian number but there is an algorithm to generate two independent Gaussian random numbers (y_1 and y_2) using two independent uniform random numbers (x_1 and x_2):

- computing $v_1 = 2x_1 - 1, v_2 = 2x_2 - 1$ and $s = v_1^2 + v_2^2$

- if $s > 1$ start again

- $y_1 = v_1 \sqrt{(-2/s) \log s}$ and $y_2 = v_2 \sqrt{(-2/s) \log s}$

Listing 6.13: in file: `nlib.py`

```
def gauss(self,mu=0.0,sigma=1.0):
    if hasattr(self,'other') and self.other:
        this, other = self.other, None
    else:
        while True:
            v1 = self.random(-1,1)
            v2 = self.random(-1,1)
            r = v1*v1+v2*v2
            if r<1: break
        this = sqrt(-2.0*log(r)/r)*v1
```

```
            self.other = sqrt(-2.0*log(r)/r)*v1
        return mu+sigma*this
```

Note how the first time the method next is called, it generates two Gaussian numbers (*this* and *other*), stores *other*, and returns *this*. Every other time, the method next is called if *other* is stored, and it returns a number; otherwise it recomputes *this* and *other* again.

To map a random Gaussian number y with mean 0 and standard deviation 1 into another Gaussian number y' with mean μ and standard deviation σ,

$$y' = \mu + y\sigma \qquad (6.70)$$

We used this relation in the last line of the code.

This is also an important distribution, and Python has a function for it:

```
random.gauss(mu,sigma)
```

Given a Gaussian random variable with mean μ and standard deviation σ, it is often useful to know how many standard deviations a correspond to a confidence c defined as

$$c = \int_{\mu-a\sigma}^{\mu+a\sigma} p(x)\mathrm{d}x \qquad (6.71)$$

The following algorithm generates a table of a versus c given μ and σ:

Listing 6.14: in file: nlib.py

```
def confidence_intervals(mu,sigma):
    """Computes the normal confidence intervals"""
    CONFIDENCE=[
        (0.68,1.0),
        (0.80,1.281551565545),
        (0.90,1.644853626951),
        (0.95,1.959963984540),
        (0.98,2.326347874041),
        (0.99,2.575829303549),
        (0.995,2.807033768344),
        (0.998,3.090232306168),
        (0.999,3.290526731492),
        (0.9999,3.890591886413),
        (0.99999,4.417173413469)
        ]
    return [(a,mu-b*sigma,mu+b*sigma) for (a,b) in CONFIDENCE]
```

6.7.4 Pareto distribution

The Pareto distribution, named after the economist Vilfredo Pareto, is a power law probability distribution that coincides with social, scientific, geophysical, actuarial, and many other types of observable phenomena. Outside the field of economics, it is sometimes referred to as the Bradford distribution. Its cumulative distribution function is

$$F(x) \equiv \mathrm{Prob}(X < x) = 1 - \left(\frac{x_m}{x}\right)^{\alpha} \tag{6.72}$$

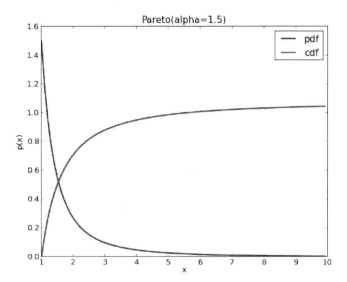

Figure 6.5: Example of Pareto distribution.

It can be implemented as follows using the inversion method:

Listing 6.15: in file: `nlib.py`

```
def pareto(self,alpha,xm):
    u = self.random()
    return xm*(1.0-u)**(-1.0/alpha)
```

The Python function to generate Pareto distributed random numbers is

```
xm * random.paretovariate(alpha)
```

The Pareto distribution is an example of Levy distribution. The Central Limit theorem does not apply to it.

6.7.5   In and on a circle

We can generate a random point (x, y) uniformly distributed on a circle by generating a random angle.

$$x = \cos(2\pi u) \tag{6.73}$$

$$y = \sin(2\pi u) \tag{6.74}$$

Listing 6.16: in file: `nlib.py`

```
def point_on_circle(self, radius=1.0):
    angle = 2.0*pi*self.random()
    return radius*math.cos(angle), radius*math.sin(angle)
```

We can generate a random point uniformly distributed inside a circle by generating, independently, the x and y coordinates of points inside a square and rejecting those outside the circle:

Listing 6.17: in file: `nlib.py`

```
def point_in_circle(self,radius=1.0):
    while True:
        x = self.uniform(-radius,radius)
        y = self.uniform(-radius,radius)
        if x*x+y*y < radius*radius:
            return x,y
```

6.7.6   In and on a sphere

A random point (x, y, z) uniformly distributed on a sphere of radius 1 is obtained by generating three uniform random numbers $u_1, u_2, u_3,$; compute $v_i = 2u_i - 1$, and if $v_1^2 + v_2^2 + v_3^2 \leq 1$,

$$x = v_1 / \sqrt{v_1^2 + v_2^2 + v_3^2} \tag{6.75}$$

$$y = v_2 / \sqrt{v_1^2 + v_2^2 + v_3^2} \tag{6.76}$$

$$z = v_3 / \sqrt{v_1^2 + v_2^2 + v_3^2} \tag{6.77}$$

else start again.

Listing 6.18: in file: `nlib.py`

```python
def point_in_sphere(self,radius=1.0):
    while True:
        x = self.uniform(-radius,radius)
        y = self.uniform(-radius,radius)
        z = self.uniform(-radius,radius)
        if x*x+y*y+z*z < radius*radius:
            return x,y,z

def point_on_sphere(self, radius=1.0):
    x,y,z = self.point_in_sphere(radius)
    norm = math.sqrt(x*x+y*y+z*z)
    return x/norm,y/norm,z/norm
```

6.8 Resampling

So far we always generated random numbers by modeling the random variable (e.g., uniform, or exponential, or Pareto) and using an algorithm to generate possible values of the random variables.

We now introduce a different methodology, which we will need later when talking about the bootstrap method. If we have a population S of equally distributed events and we need to generate an event from the same distribution as the population, we can simply draw a random element from the population. In Python, this is done with

```python
>>> S = [1,2,3,4,5,6]
>>> print random.choice(S)
```

We can therefore generate a sample of random events by repeating this procedure. This is called *resampling* [53]:

Listing 6.19: in file: `nlib.py`

```python
def resample(S,size=None):
    return [random.choice(S) for i in xrange(size or len(S))]
```

6.9   Binning

Binning is the process of dividing a space of possible events into partitions and counting how many events fall into each partition. We can bin the numbers generated by a pseudo random number generator and measure the distribution of the random numbers.

Let's consider the following program:

```
def bin(generator,nevents,a,b,nbins):
    # create empty bins
    bins=[]
    for k in xrange(nbins):
        bins.append(0)
    # fill the bins
    for i in xrange(nevents):
        x=generator.uniform()
        if x>=a and x<=b:
            k=int((x-a)/(b-a)*nbins)
            bins[k]=bins[k]+1
    # normalize bins
    for i in xrange(nbins):
        bins[i]=float(bins[i])/nevents
    return bins

def test_bin(nevents=1000,nbins=10):
    bins=bin(MCG(time()),nevents,0,1,nbins)
    for i in xrange(len(bins)):
        print i, bins[i]

>>> test_bin()
```

It produces the following output:

```
i frequency[i]
0 0.101
1 0.117
2 0.092
3 0.091
4 0.091
5 0.122
6 0.096
7 0.102
8 0.090
9 0.098
```

Note that

- all bins have the same size 1/nbins;

- the size of the bins is normalized, and the sum of the values is 1

- the distribution of the events into bins approaches the distribution of the numbers generated by the random number generator

As an experiment, we can do the same binning on a larger number of events,

```
>>> test_bin(100000)
```

which produces the following output:

```
i frequency[i]
0 0.09926
1 0.09772
2 0.10061
3 0.09894
4 0.10097
5 0.09997
6 0.10056
7 0.09976
8 0.10201
9 0.10020
```

Note that these frequencies differ from 0.1 for less than 3%, whereas some of the preceding numbers differ from 0.11 for more than 20%.

7

Monte Carlo Simulations

7.1 Introduction

Monte Carlo methods are a class of algorithms that rely on repeated random sampling to compute their results, which are otherwise deterministic.

7.1.1 Computing π

The standard way to compute π is by applying the definition: π is the length of a semicircle with a radius equal to 1. From the definition, one can derive an exact formula:

$$\pi = 4 \arctan 1 \tag{7.1}$$

The arctan has the following Taylor series expansion:[1]:

$$\arctan x = \sum_{i=0}^{\infty} (-1)^i \frac{x^{2i+1}}{2i+1} \tag{7.8}$$

---

[1]Taylor expansion:

$$f(x) = f(0) + f'(0)x + \frac{1}{2!}f''(0)x^2 + \dots + \frac{1}{i!}f^{(i)}(0)x^2 + \dots \tag{7.2}$$

and one can approximate π to arbitrary precision by computing the sum

$$\pi = \sum_{i=0} (-1)^i \frac{4}{2i+1} \qquad (7.9)$$

We can use the program

```
def pi_Taylor(n):
    pi=0
    for i in xrange(n):
        pi=pi+4.0/(2*i+1)*(-1)**i
        print i,pi

>>> pi_Taylor(1000)
```

which produces the following output:

```
0 4.0
1 2.66666...
2 3.46666...
3 2.89523...
4 3.33968...
...
999 3.14..
```

A better formula is due to Plauffe,

$$\pi = \sum_{i=0} \frac{1}{16^i} \left(\frac{4}{8i+1} - \frac{2}{8i+4} - \frac{1}{8i+5} - \frac{1}{8i+6} \right) \qquad (7.10)$$

which we can implement as follows: we can use the program

```
from decimal import Decimal
def pi_Plauffe(n):
    pi=Decimal(0)
```

---

and if $f(x) = \arctan x$ then:

$$f'(x) \quad = \quad \frac{d \arctan x}{dx} = \frac{1}{1+x^2} \to f'(0) = 1 \qquad (7.3)$$

$$f''(x) \quad = \quad \frac{d^2 \arctan x}{d^2 x} = \frac{d}{dx}\frac{1}{1+x^2} = -\frac{2x}{(1+x^2)^2} \qquad (7.4)$$

$$\ldots \qquad (7.5)$$

$$f^{(2i+1)}(x) \quad = \quad (-1)^i \frac{(2i)!}{(1+x^2)^{2i+1}} + x\ldots \to f^{(2i+1)}(0) = (-1)(2i)! \qquad (7.6)$$

$$f^{(2i)}(x) \quad = \quad x\ldots \to f^{(2i)}(0) = 0 \qquad (7.7)$$

```
4    a,b,c,d = Decimal(4),Decimal(2),Decimal(1),Decimal(1)/Decimal(16)
5    for i in xrange(n):
6        i8 = Decimal(8)*i
7        pi=pi+(d**i)*(a/(i8+1)-b/(i8+4)-c/(i8+5)-c/(i8+6))
8    return pi
9 >>> pi_Plauffe(1000)
```

The preceding formula works and converges very fast and already in 100 iterations produces

$$\pi = 3.14159265358979932384626433... \tag{7.11}$$

There is a different approach based on the fact that π is also the area of a circle of radius 1. We can draw a square or area containing a quarter of a circle of radius 1. We can randomly generate points (x, y) with uniform distribution inside the square and check if the points fall inside the circle. The ratio between the number of points that fall in the circle over the total number of points is proportional to the area of the quarter of a circle ($\pi/4$) divided by the area of the square (1).

Here is a program that implements this strategy:

```
1  from random import *
2
3  def pi_mc(n):
4      pi=0
5      counter=0
6      for i in xrange(n):
7          x=random()
8          y=random()
9          if x**2 + y**2 < 1:
10             counter=counter+1
11         pi=4.0*float(counter)/(i+1)
12         print i,pi
13
14 pi_mc(1000)
```

The output of the algorithm is shown in fig. 7.1.1.

The convergence rate in this case is very slow, and this algorithm is of no practical value, but the methodology is sound, and for some problems, this method is the only one feasible.

Let's summarize what we have done: we have formulated our problem (compute π) as the problem of computing an area (the area of a quarter

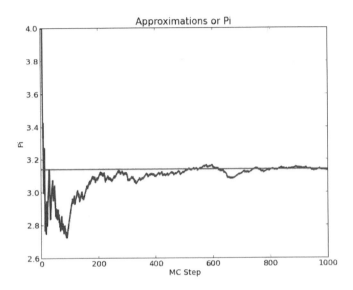

Figure 7.1: Convergence of **pi_mc**.

of a circle), and we have computed the area using random numbers. This is a particular example of a more general technique known as a Monte Carlo integration. In fact, the computation of an area is equivalent to the problem of computing an integral.

Sometimes the formula is not known, or it is too complex to compute reliably, hence a Monte Carlo solution becomes preferable.

7.1.2 Simulating an online merchant

Let's consider an online merchant. A website is visited many times a day. From the logfile of the web application, we determine that the average number of visitors in a day is 976, the number of visitors is Gaussian distributed, and the standard deviation is 352. We also observe that each visitor has a 5% probability of purchasing an item if the item is in stock and a 2% probability to buy an item if the item is not in stock.

The merchant sells only one type of item that costs $100 per unit. The merchant maintains N items in stock. The merchant pays $30 a day to store each unit item in stock. What is the optimal N to maximize the average daily income of the merchant?

This problem cannot easily be formulated analytically or reduced to the computation of an integral, but it can easily be simulated.

In particular, we simulate many days and, for each day i, we start with N items in stock, and we loop over each simulated visitor. If the visitor finds an item in stock, he buys it with a 5% probability (producing an income of $70), whereas if the item is not in stock, he buys it with 2% probability (producing an income of $100). At the end of each day, we pay $30 for each item remaining in stock.

Here is a program that takes N (the number of items in stock) and d (the number of simulated days) and computes the average daily income:

```
 1  def simulate_once(N):
 2      profit = 0
 3      loss = 30*N
 4      instock = N
 5      for j in xrange(int(gauss(976,352))):
 6          if instock>0:
 7              if random()<0.05:
 8                  instock=instock-1
 9                  profit = profit + 100
10          else:
11              if random()<0.02:
12                  profit = profit + 100
13      return profit-loss
14
15  def simulate_many(N,ap=1,rp=0.01,ns=1000):
16      s = 0.0
17      for k in xrange(1,ns):
18          x = simulate_once(N)
19          s += x
20          mu = s/k
21          if k>10 and mu-mu_old<max(ap,rp*mu):
22              return mu
23          else:
24              mu_old = mu
25      raise ArithmeticError('no convergence')
```

By looping over different N (items in stock), we can compute the average daily income as a function of N:

```
1 >>> for N in xrange(0,100,10):
2 >>>     print N,simulate_many(N,ap=100)
```

The program produces the following output:

```
1  n income
2  0 1955
3  10 2220
4  20 2529
5  30 2736
6  40 2838
7  50 2975
8  60 2944
9  70 2711
10 80 2327
11 90 2178
```

From this we deduce that the optimal number of items to carry in stock is about 50. We could increase the resolution and precision of the simulation by increasing the number of simulated days and reducing the step of the amount of items in stock.

Note that the statement gauss(976,352) generates a random floating point number with a Gaussian distribution centered at 976 and standard deviation equal to 352, whereas the statement

```
1 if random()<0.05:
```

ensures that the subsequent block is executed with a probability of 5%.

The basic ingredients of every Monte Carlo simulation are here: (1) a function that simulates the system once and uses random variables to model unknown quantities; (2) a function that repeats the simulation many times to compute an average.

Any Monte Carlo solver comprises the following parts:

- A generator of random numbers (such as we have discussed in the previous chapter)

- A function that uses the random number generator and can simulate the system once (we will call x the result of each simulate once)

- A function that calls the preceding simulation repeatedly and averages the results until they converge $\mu = \frac{1}{N} \sum x_i$

- A function to estimate the accuracy of the result and determine when to stop the simulation, $\delta\mu < $ precision

7.2   Error analysis and the bootstrap method

The result of any MC computation is an average:

$$\mu = \frac{1}{N}\sum x_i \qquad\qquad (7.12)$$

The error on this average can be estimated using the formula

$$\delta\mu = \frac{\sigma}{\sqrt{N}} = \sqrt{\frac{1}{N}\left(\frac{1}{N}\sum x_i^2 - \mu^2\right)} \qquad\qquad (7.13)$$

This formula assumes the distribution of the x_i is Gaussian. Using this formula, we can compute a 68% confidence level for the MC computation of π, shown in fig. 7.2.

The purpose of the bootstrap [54] algorithm is computing the error in an average $\mu = (1/N)\sum x_i$ without making the assumption that the x_i are Gaussian.

The first step of the bootstrap methodology consists of computing the average not only on the initial sample $\{x_i\}$ but also on many data samples obtained by resampling the original data. If the number of elements N of the original sample were infinity, the average on each other sample would be the same. Because N is finite, each of these means produces slightly different results:

$$\mu_k = \frac{1}{N}\sum_i x_i^{[k]} \qquad\qquad (7.14)$$

where $x_i^{[k]}$ is the ith element of resample k and μ_k is the average of that resample.

The second step of the bootstrap methodology consists of sorting the μ_k and finding two values μ^- and μ^+ that with a given percentage of the

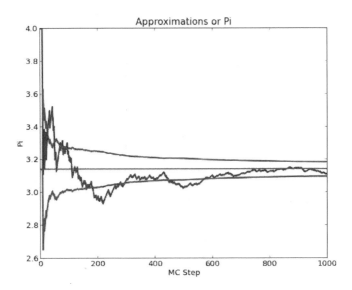

Figure 7.2: Convergence of π.

means follows in between those two values. The given percentage is the confidence level, and we set it to 68%.

Here is the complete algorithm:

Listing 7.1: in file: `nlib.py`

```
def bootstrap(x, confidence=0.68, nsamples=100):
    """Computes the bootstrap errors of the input list."""
    def mean(S): return float(sum(x for x in S))/len(S)
    means = [mean(resample(x)) for k in xrange(nsamples)]
    means.sort()
    left_tail = int(((1.0-confidence)/2)*nsamples)
    right_tail = nsamples-1-left_tail
    return means[left_tail], mean(x), means[right_tail]
```

Here is an example of usage:

```
>>> S = [random.gauss(2,1) for k in range(100)]
>>> print bootstrap(S)
(1.7767055865879007, 1.8968778392283303, 2.003420362236985)
```

In this example, the output consists of μ^-, μ, and μ^+.

Because S contains 100 random Gaussian numbers, with average 2 and standard deviation 1, we expect μ to be close to 2. We get 1.89. The bootstrap tells us that with 68% probability, the true average of these numbers is indeed between 1.77 and 2.00. The uncertainty $(2.00 - 1.77)/2 = 0.12$ is compatible with $\sigma/\sqrt{100} = 1/10 = 0.10$.

7.3 A general purpose Monte Carlo engine

We can now combine everything we have seen so far into a generic program that can be used to perform the most generic Monte Carlo computation/simulation:

Listing 7.2: in file: `nlib.py`

```python
class MCEngine:
    """
    Monte Carlo Engine parent class.
    Runs a simulation many times and computes average and error in average.
    Must be extended to provide the simulate_once method
    """
    def simulate_once(self):
        raise NotImplementedError

    def simulate_many(self, ap=0.1, rp=0.1, ns=1000):
        self.results = []
        s1=s2=0.0
        self.convergence=False
        for k in xrange(1,ns):
            x = self.simulate_once()
            self.results.append(x)
            s1 += x
            s2 += x*x
            mu = float(s1)/k
            variance = float(s2)/k-mu*mu
            dmu = sqrt(variance/k)
            if k>10:
                if abs(dmu)<max(ap,abs(mu)*rp):
                    self.converence = True
                    break
        self.results.sort()
        return bootstrap(self.results)
```

The preceding class has two methods:

- `simulate_once` is not implemented because the class is designed to be subclassed, and the method is supposed to be implemented for each specific computation.

- `simulate_many` is the part that stays the same; it calls `simulate_once` repeatedly, computes average and error analysis, checks convergence, and computes bootstrap error for the result.

It is also useful to have a function, which we call **var** (aka *value at risk* [55]), which computes a numerical value so that the output of a given percentage of the simulations falls below that value:

Listing 7.3: in file: `nlib.py`

```
def var(self, confidence=95):
    index = int(0.01*len(self.results)*confidence+0.999)
    if len(self.results)-index < 5:
        raise ArithmeticError('not enough data, not reliable')
    return self.results[index]
```

Now, as a first example, we can recompute π using this class:

```
>>> class PiSimulator(MCEngine):
...     def simulate_once(self):
...         return 4.0 if (random.random()**2+random.random()**2)<1 else 0.0
...
>>> s = PiSimulator()
>>> print s.simulate_many()
(2.1818181818181817, 2.909090909090909, 3.6363636363636362)
```

Our engine finds that the value of π with 68% confidence level is between 2.18 and 3.63, with the most likely value of 2.90. Of course, this is incorrect, because it generates too few samples, but the bounds are correct, and that is what matters.

7.3.1 Value at risk

Let's consider a business subject to random losses, for example, a large bank subject to theft from employees. Here we will make the following reasonable assumptions (which have been verified with data):

- There is no correlation between individual events.

- There is no correlation between the time when a loss event occurs and the amount of the loss.

- The time interval between losses is given by the exponential distribution (this is a Poisson process).

- The distribution of the loss amount is a Pareto distribution (there is a fat tail for large losses).

- The average number of losses is 10 per day.

- The minimum recorded loss is $5000. The average loss is $15,000.

Our goal is to simulate one year of losses and to determine

- The average total yearly loss

- How much to save to make sure that in 95% of the simulated scenarios, the losses can be covered without going broke

From these assumptions, we determine that the $\lambda = 10$ for the exponential distribution and $x_m = 3000$ for the Pareto distribution. The mean of the Pareto distribution is $\alpha x_m / (\alpha - 1) = 15,000$, from which we determine that $\alpha = 1.5$.

We can answer the first questions (the average total loss) simply multiplying the average number of losses per year, 52×5, by the number of losses in one day, 10, and by the average individual loss, $15,000, thus obtaining

$$[\text{average yearly loss}] = \$39,000,000 \qquad (7.15)$$

To answer the second question, we would need to study the width of the distribution. The problem is that, for $\alpha = 1.5$, the standard deviation of the Pareto distribution is infinity, and analytical methods do not apply. We can do it using a Monte Carlo simulation:

Listing 7.4: in file: risk.py

```
from nlib import *
import random

class RiskEngine(MCEngine):
    def __init__(self, lamb, xm, alpha):
        self.lamb = lamb
```

```
7        self.xm = xm
8        self.alpha = alpha
9    def simulate_once(self):
10       total_loss = 0.0
11       t = 0.0
12       while t<260:
13           dt = random.expovariate(self.lamb)
14           amount = self.xm*random.paretovariate(self.alpha)
15           t += dt
16           total_loss += amount
17       return total_loss
18
19 def main():
20     s = RiskEngine(lamb=10, xm=5000, alpha=1.5)
21     print s.simulate_many(rp=1e-4,ns=1000)
22     print s.var(95)
23
24 main()
```

This produces the following output:

```
1 (38740147.179054834, 38896608.25084647, 39057683.35621854)
2 45705881.8776
```

The output of simulate_many should be compatible with the true result (defined as the result after an infinite number of iterations and at infinite precision) within the estimated statistical error.

The output of the var function answers our second questions: We have to save $45,705,881 to make sure that in 95% of cases our losses are covered by the savings.

7.3.2 Network reliability

Let's consider a network represented by a set of n_{nodes} nodes and n_{links} bidirectional links. Information packets travel on the network. They can originate at any node (start) and be addressed to any other node (stop). Each link of the network has a probability p of transmitting the packet (success) and a probability $(1 - p)$ of dropping the packet (failure). The probability p is in general different for each link of the network.

We want to compute the probability that a packet starting in start finds a successful path to reach stop. A path is successful if, for a given simulation, all links in the path succeed in carrying the packet.

The key trick in solving this problem is in finding the proper representation for the network. Since we are not requiring to determine the exact path but only proof of existence, we use the concept of equivalence classes.

We say that two nodes are in the same equivalence class if and only if there is a successful path that connects the two nodes.

The optimal data structure to implement equivalence classes is DisjSets, discussed in chapter 3.

To simulate the system, we create a class Network that extends MCEngine. It has a simulate_once method that tries to send a packet from start to stop and simulates the network once. During the simulation each link of the network may be up or down with given probability. If there is a path connecting the start node to the stop node in which all links of the network are up, than the packet transfer succeeds. We use the DisjointSets to represent sets of nodes connected together. If there is a link up connecting a node from a set to a node in another set, than the two sets are joined. If, in the end, the start and stop nodes are found to belong to the same set, then there is a path and simulate_once returns 1, otherwise it returns 0.

Listing 7.5: in file: network.py

```
from nlib import *
import random

class NetworkReliability(MCEngine):
    def __init__(self,n_nodes,start,stop):
        self.links = []
        self.n_nodes = n_nodes
        self.start = start
        self.stop = stop
    def add_link(self,i,j,failure_probability):
        self.links.append((i,j,failure_probability))
    def simulate_once(self):
        nodes = DisjointSets(self.n_nodes)
        for i,j,pf in self.links:
            if random.random()>pf:
                nodes.join(i,j)
        return nodes.joined(i,j)

def main():
```

```
20    s = NetworkReliability(100,start=0,stop=1)
21    for k in range(300):
22        s.add_link(random.randint(0,99),
23                   random.randint(0,99),
24                   random.random())
25    print s.simulate_many()
26
27 main()
```

7.3.3 Critical mass

Here we consider the simulation of a chain reaction in a fissile material, for example, the uranium in a nuclear reactor [56]. We assume a material is in a spherical shape of known radius. At each point there is a probability of a nuclear fission, which we model as the emission of two neutrons. Each of the two neutrons travels and hits an atom, thus causing another fission. The two neutrons are emitted in random opposite directions and travel a distance given by the exponential distribution. The new fissions may occur inside material itself or outside. If outside, they are ignored. If the number of fission events inside the material grows exponentially with time, we have a self-sustained chain reaction; otherwise, we do not.

Fig. 7.3.3 provides a representation of the process.

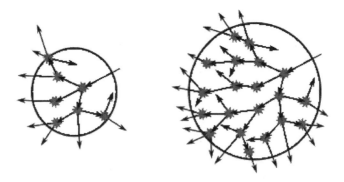

Figure 7.3: Example of chain reaction within a fissile material. If the mass is small, most of the decay products escape (left, sub-criticality), whereas if the mass exceeds a certain critical mass, there is a self-sustained chain reaction (right).

Here is a possible implementation of the process. We store each event

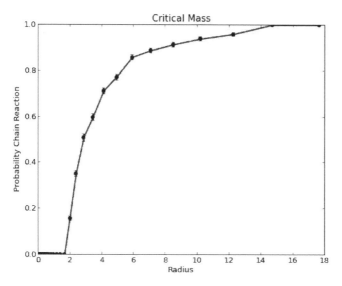

Figure 7.4: Probability of chain reaction in uranium.

that happens inside the material in a queue (events). For each simulation, the queue starts with one event, and from it we generate two more, and so on. If the new events happen inside the material, we place the new events back in the queue. If the size of the queue shrinks to zero, then we are subcritical. If the size of the queue grows exponentially, we have a self-sustained chain reaction. We detect this by measuring the size of the queue and wether it exceeds a threshold (which we arbitrarily set to 200). The average free flight distance for a neutron in uranium is 1.91 cm. We use this number in our simulation. Given the radius of the material, simulate_once returns 1.0 if it detects a chain reaction and 0.0 if it does not. The output of simulate_many is the probability of a chain reaction:

Listing 7.6: in file: nuclear.py

```
from nlib import *
import math
import random

class NuclearReactor(MCEngine):
    def __init__(self,radius,mean_free_path=1.91,threshold=200):
```

```
 7        self.radius = radius
 8        self.density = 1.0/mean_free_path
 9        self.threshold = threshold
10    def point_on_sphere(self):
11        while True:
12            x,y,z = random.random(), random.random(), random.random()
13            d = math.sqrt(x*x+y*y+z*z)
14            if d<1: return (x/d,y/d,z/d) # project on surface
15    def simulate_once(self):
16        p = (0,0,0)
17        events = [p]
18        while events:
19            event = events.pop()
20            v = self.point_on_sphere()
21            d1 = random.expovariate(self.density)
22            d2 = random.expovariate(self.density)
23            p1 = (p[0]+v[0]*d1,p[1]+v[1]*d1,p[2]+v[2]*d1)
24            p2 = (p[0]-v[0]*d2,p[1]-v[1]*d2,p[2]-v[2]*d2)
25            if p1[0]**2+p1[1]**2+p1[2]**2 < self.radius:
26                events.append(p1)
27            if p2[0]**2+p2[1]**2+p2[2]**2 < self.radius:
28                events.append(p2)
29            if len(events) > self.threshold:
30                return 1.0
31        return 0.0
32
33 def main():
34    s = NuclearReactor(MCEngine)
35    data = []
36    s.radius = 0.01
37    while s.radius<21:
38        r = s.simulate_many(ap=0.01,rp=0.01,ns=1000,nm=100)
39        data.append((s.radius, r[1], (r[2]-r[0])/2))
40        s.radius *= 1.2
41    c = Canvas(title='Critical Mass',xlab='Radius',ylab='Probability Chain
             Reaction')
42    c.plot(data).errorbar(data).save('nuclear.png')
43
44 main()
```

Fig. 7.3.3 shows the output of the program, the probability of a chain reaction as function of the size of the uranium mass. We find a critical radius between 2 cm and 10 cm, which corresponds to a critical mass between 0.5 kg and 60 kg. The official number is 15 kg for uranium 233 and 60 kg for uranium 235. The lesson to learn here is that it is not safe to accumulate too much fissile material together. This simulation can be easily tweaked to determine the thickness of a container required to shield a radioactive material.

7.4 Monte Carlo integration

7.4.1 One-dimensional Monte Carlo integration

Let's consider a one-dimensional integral

$$I = \int_a^b f(x)dx \tag{7.16}$$

Let's now determine two functions $g(x)$ and $p(x)$ such that

$$p(x) = 0 \text{ for } x \in [-\infty, a] \cup [n, \infty] \tag{7.17}$$

and

$$\int_{-\infty}^{+\infty} p(x)dx = 1 \tag{7.18}$$

and

$$g(x) = f(x)/p(x) \tag{7.19}$$

We can interpret $p(x)$ as a probability mass function and

$$E[g(X)] = \int_{-\infty}^{+\infty} g(x)p(x)dx = \int_a^b f(x)dx = I \tag{7.20}$$

Therefore we can compute the integral by computing the expectation value of the function $g(X)$, where X is a random variable with a distribution (probability mass function) $p(x)$ different from zero in $[a, b]$ generated.

An obvious, although not in general an optimal choice, is

$$p(x) \equiv \left\{ \begin{array}{ll} 1/(b-a) & \text{if } x \in [a, b] \\ 0 & \text{otherwise} \end{array} \right\} \tag{7.21}$$

$$g(x) \equiv (b-a)f(x)$$

so that X is just a uniform random variable in $[a, b]$. Therefore

$$I = E\left[g(X)\right] = \frac{1}{N} \sum_{i=0}^{i<N} g(x_i) \tag{7.22}$$

This means that the integral can be evaluated by generating N random points x_i with uniform distribution in the domain, evaluating the integrand (the function f) on each point, averaging the results, and multiplying the average by the size of the domain $(b - a)$.

Naively, the error on the result can be estimated by computing the variance

$$\sigma^2 = \frac{1}{N} \sum_{i=0}^{i<N} [g(x_i) - \langle g \rangle]^2 \tag{7.23}$$

with

$$\langle g \rangle = \frac{1}{N} \sum_{i=0}^{i<N} g(x_i) \tag{7.24}$$

and the error on the result is given by:

$$\delta I = \sqrt{\frac{\sigma^2}{N}} \tag{7.25}$$

The larger the set of sample points N, the lower the variance and the error. The larger N, the better $E[g(X)]$ approximates the correct result I.

Here is a program in Python:

Listing 7.7: in file: `integrate.py`

```
class MCIntegrator(MCEngine):
    def __init__(self,f,a,b):
        self.f = f
        self.a = a
        self.b = b
    def simulate_once(self):
        a, b, f = self.a, self.b, self.f
        x = a+(b-a)*random.random()
        g = (b-a)*f(x)
        return g

def main():
    s = MCIntegrator(f lambda x: math.sin(x),a=0,b=1)
    print s.simulate_many()

main()
```

This technique is very general and can be extended to almost any integral assuming the integrand is smooth enough on the integration domain.

The choice (7.21) is not always optimal because the integrand may be very small in some regions of the integration domain and very large in other regions. Clearly some regions contribute more than others to the average, and one would like to generate points with a probability mass function that is as close as possible to the original integrand. Therefore we should choose a $p(x)$ according to the following conditions:

- $p(x)$ is very similar and proportional to $f(x)$

- given $F(x) = \int_{-\infty}^{x} p(x)dx$, $F^{-1}(x)$ can be computed analytically.

Any choice for $p(x)$ that makes the integration algorithm converge faster with less calls to simulate_once is called a *variance reduction technique*.

7.4.2 Two-dimensional Monte Carlo integration

The technique described earlier can easily be extended to two-dimensional integrals:

$$I = \int_{\mathfrak{D}} f(x_0, x_1)dx_0dx_1 \qquad (7.26)$$

where $\mathfrak{D}$ is some two-dimensional domain. We determine two functions $g(x_0, x_1)$ and $p_0(x_0), p_1(x_1)$ such that

$$p_0(x_0) = 0 \text{ or } p_1(x_1) = 0 \text{ for } x \notin \mathfrak{D} \qquad (7.27)$$

and

$$\int p_0(x_0)p_1(x_1)dx_0dx_1 = 1 \qquad (7.28)$$

and

$$g(x_0, x_1) = \frac{f(x_0, x_1)}{p_0(x_0)p_1(x_1)} \qquad (7.29)$$

We can interpret $p(x_0, x_1)$ as a probability mass function for two independent random variables X_0 and X_1 and

$$E[g(X_0, X_1)] = \int g(x_0, x_1)p_0(x_0)p_1(x_1)dx = \int_{\mathfrak{D}} f(x_0, x_1)dx_0dx_1 = I$$
$$(7.30)$$

Therefore

$$I = E[g(X_0, X_1)] = \frac{1}{N} \sum_{i=0}^{i<N} g(x_{i0}, x_{i1}) \qquad (7.31)$$

7.4.3 n-dimensional Monte Carlo integration

The technique described earlier can also be extended to n-dimensional integrals

$$I = \int_{\mathfrak{D}} f(x_0, ..., x_{n-1}) dx_0 ... dx_{n-1} \qquad (7.32)$$

where $\mathfrak{D}$ is some n-dimensional domain identified by a function $domain(x_0, ..., x_{n-1})$ equal to 1 if $\mathbf{x} = (x_0, ..., x_{n-1})$ is in the domain, o otherwise. We determine two functions $g(x_0, ..., x_{n-1})$ and $p(x_0, ..., x_{n-1})$ such that

$$p(x_0, ..., x_{n-1}) = 0 \text{ for } x \notin \mathfrak{D} \qquad (7.33)$$

and

$$\int p(x_0, ..., x_{n-1}) dx_0 ... dx_{n-1} = 1 \qquad (7.34)$$

and

$$g(x_0, ..., x_{n-1}) = f(x_0, ..., x_{n-1}) / p(x_0, ..., x_{n-1}) \qquad (7.35)$$

We can interpret $p(x_0, ..., x_{n-1})$ as a probability mass function for n independent random variables $X_0 ... X_{n-1}$ and

$$E[g(X_0, ..., X_{n-1})] = \int g(x_0, ..., x_{n-1}) p(x_0, ..., x_{n-1}) dx \qquad (7.36)$$

$$= \int_{\mathfrak{D}} f(x_0, ..., x_{n-1}) dx_0 ... dx_{n-1} = I \qquad (7.37)$$

Therefore

$$I = E[g(X_0, .., X_{n-1})] = \frac{1}{N} \sum_{i=0}^{i<N} g(\mathbf{x}_i) \qquad (7.38)$$

where for every point $\mathbf{x}_i$ is a tuple $(x_{i0}, x_{i1}, ..., x_{i,n-1})$.

As an example, we consider the integral

$$I = \int_0^1 dx_0 \int_0^1 dx_1 \int_0^1 dx_2 \int_0^1 dx_3 \sin(x_0 + x_1 + x_2 + x_3) \qquad (7.39)$$

Here is the Python code:

```python
>>> class MCIntegrator(MCEngine):
...     def simulate_once(self):
...         volume = 1.0
...         while True:
...             x = [random.random() for d in range(4)]
...             if sum(xi**2 for xi in x)<1: break
...         return volume*self.f(x)
>>> s = MCIntegrator()
>>> s.f = lambda x: math.sin(x[0]+x[1]+x[2]+x[3])
>>> print s.simulate_many()
```

7.5   Stochastic, Markov, Wiener, and processes

A *stochastic process* [57] is a random function, for example, a function that
maps a variable n with domain D into X_n, where X_n is a random variable with domain R. In practical applications, the domain D over which
the function is defined can be a time interval (and the stochastic is called
a *time series*) or a region of space (and the stochastic process is called a
random field). Familiar examples of time series include *random walks* [58];
stock market and exchange rate fluctuations; signals such as speech, audio, and video; or medical data such as a patient's EKG, EEG, blood
pressure, or temperature. Examples of random fields include static images, random topographies (landscapes), or composition variations of an
inhomogeneous material.

Let's consider a grasshopper moving on a straight line, and let X_n be the
position of the grasshopper at time $t = n\Delta_t$. Let's also assume that at time
0, $X_0 = 0$. The position of the grasshopper at each future ($t > 0$) time is
unknown. Therefore it is a random variable.

We can model the movements of the grasshopper as follows:

$$X_{n+1} = X_n + \mu + \varepsilon_n\Delta_x \tag{7.40}$$

where Δ_x is a fixed step and ε_n is a random variable whose distribution
depends on the model; μ is a constant drift term (think of wind pushing
the grasshopper in one direction). It is clear that X_{n+1} only depends on

304   ANNOTATED ALGORITHMS IN PYTHON

X_n and ε_n; therefore the probability distribution of X_{n+1} only depends on X_n and the probability distribution of ε_n, but it does not depend on the past history of the grasshopper's movements at times $t < n\Delta_t$. We can write the statement by saying that

$$\text{Prob}(X_{n+1} = x | \{X_i\} \text{ for } i \leq n) = \text{Prob}(X_{n+1} = x | X_n) \qquad (7.41)$$

A process in which the probability distribution of its future state only depends on the present state and not on the past is called a *Markov process* [59].

To complete our model, we need to make additional assumptions about the probability distribution of ε_n. We consider the two following cases:

- ε_n is a random variable with a Bernoulli distribution ($\varepsilon_n = +1$ with probability p and $\varepsilon_n = -1$ with probability $1 - p$).

- ε_n is a random variable with a normal (Gaussian) distribution with probability mass function $p(\varepsilon) = e^{-\varepsilon^2/2}$. Notice that the previous case (Bernoulli) is equivalent to this case (Gaussian) over long time intervals because the sum of many independent Bernoulli variables approaches a Gaussian distribution.

A continuous time stochastic process (when ε_n is a continuous random number) is called a *Wiener process* [60].

The specific case when ε_n is a Gaussian random variable is called an *Ito process* [61]. An Ito process is also a Wiener process.

7.5.1   Discrete random walk (Bernoulli process)

Here we assume a discrete random walk: ε_n equal to $+1$ with probability p and equal to -1 with probability $1 - p$. We consider discrete time intervals of equal length Δ_t; at each time step, if $\varepsilon_n = +1$, the grasshopper moves forward one unit (Δ_x) with probability p, and if $\varepsilon_n = -1$, he moves backward one unit ($-\Delta_x$) with probability $1 - p$.

For a total n steps, the probability of moving n_+ steps in a positive direction and $n_- = n - n_+$ in a negative direction is given by

$$\frac{n!}{n_+!(n - n_+)!} p^{n_+} (1 - p)^{n - n_+} \qquad (7.42)$$

The probability of going from $a = 0$ to $b = k\Delta_x > 0$ in a time $t = n\Delta_t > 0$ corresponds to the case when

$$n = n_+ + n_i \qquad (7.43)$$
$$k = n_+ - n_- \qquad (7.44)$$

that solved in n_+ gives $n_+ = (n + k)/2$, and therefore the probability of going from 0 to k in time $t = n\Delta_t$ is given by

$$\text{Prob}(n, k) = \frac{n!}{((n + k)/2)!((n - k)/2)!} p^{(n+k)/2} (1 - p)^{(n-k)/2} \qquad (7.45)$$

Note that $n + k$ has to be even, otherwise it is not possible for the grasshopper to reach $k\Delta_x$ in exactly n steps.

For large n, the following distribution in k/n tends to a Gaussian distribution.

7.5.2 Random walk: Ito process

Let's assume an Ito process for our random walk: ε_n is normally (Gaussian) distributed. We consider discrete time intervals of equal length Δ_t, at each time step if $\varepsilon_n = \varepsilon$ with probability mass function $p(\varepsilon) = e^{-\varepsilon^2/2}$. It turns out that eq.(7.40) gives

$$X_n = n\mu + \Delta_x \sum_{i=0}^{i<n} \varepsilon_i \qquad (7.46)$$

Therefore the location of the random walker at time $t = n\Delta_t$ is given by the sum of n normal (Gaussian) random variables:

$$p(X_n) = \frac{1}{\sqrt{2\pi n \Delta_x^2}} e^{-(X_n - n\mu)^2/(2n\Delta_x^2)} \qquad (7.47)$$

Notice how the mean and the variance of X_n are both proportional to n, whereas the standard deviation is proportional to $\sqrt{n}$.

$$\text{Prob}(a \leq X_n \leq b) = \frac{1}{\sqrt{2\pi n \Delta_x^2}} \int_a^b e^{-(X_n - n\mu)^2/(2n\Delta_x^2)} dx \quad (7.48)$$

$$= \frac{1}{\sqrt{2\pi}} \int_{(a-n\mu)/(\sqrt{n}\Delta_x)}^{(b-n\mu)/(\sqrt{n}\Delta_x)} e^{-x^2/2} dx \quad (7.49)$$

$$= \text{erf}(\frac{b-n\mu}{\sqrt{n}\Delta_x}) - \text{erf}(\frac{a-n\mu}{\sqrt{n}\Delta_x}) \quad (7.50)$$

7.6 Option pricing

A European call option is a contract that depends on an asset S. The contract gives the buyer of the contract the right (the option) to buy S at a fixed price A some time in the future, even if the actual price S may be different. The actual current price of the asset is called the *spot price*. The buyer of the option hopes that the price of the asset, S_t, will exceed A, so that he will be able to buy it at a discount, sell it at market price, and make a profit. The seller of the option hopes this does not happen, so he earns the full sale price. For the buyer of the option, the worst case scenario is not to be able to recover the price paid for the option, but there is no best case because, hypothetically, he can make an arbitrarily large profit. For the seller, it is the opposite. He has an unlimited liability.

In practice, a call option allows a buyer to sell risk (the risk of the price of S going up) to the seller. He pays a price for it, the cost of the option. This is a form of insurance. There are two types of people who trade options: those who are willing to pay to get rid of risk (because they need the underlying asset and want it at a guaranteed price) and those who simply speculate (buy risk and sell insurance). On average, speculators make money because, if they sell many options, risk averages out, and they collect the premiums (the cost of the options).

The European option has a term or expiration, τ. It can only be exercised at expiration. The amount A is called the *strike price*.

The value at expiration of a European call option is

$$\max(S_\tau - A, 0) \tag{7.51}$$

Its present value is therefore

$$\max(S_\tau - A, 0)e^{-r\tau} \tag{7.52}$$

where r is the risk-free interest rate. This value corresponds to how much we would have to borrow today from a bank so that we can repay the bank at time τ with the profit from the option.

All our knowledge about the future spot price $x = S_\tau$ of the underlying asset can be summarized into a probability mass function $p_\tau(x)$. Under the assumption that $p_\tau(x)$ is known to both the buyer and the seller of the option, it has to be that the averaged net present value of the option is zero for any of the two parties to want to enter into the contract. Therefore

$$C_{call} = e^{-r\tau} \int_{-\infty}^{+\infty} \max(x - A, 0) p_\tau(x) dx \tag{7.53}$$

Similarly, we can perform the same computations for a put option. A put option gives the buyer the option to sell the asset on a given day at a fixed price. This is an insurance against the price going down instead of going up. The value of this option at expiration is

$$\max(A - S_\tau, 0) \tag{7.54}$$

and its pricing formula is

$$C_{put} = e^{-r\tau} \int_{-\infty}^{+\infty} \max(A - x, 0) p_\tau(x) dx \tag{7.55}$$

Also notice that $C_{call} - C_{put} = S_0 - Ae^{-r\tau}$. This relation is called the *call-put parity*.

Our goal is to model $p_\tau(x)$, the distribution of possible prices for the underlying asset at expiration of the option, and compute the preceding integrals using Monte Carlo.

7.6.1 Pricing European options: Binomial tree

To price an option, we need to know $p_\tau(S_\tau)$. This means we need to know something about the future behavior of the price S_τ of the underlying asset S (a stock, an index, or something else). In absence of other information (crystal ball or illegal insider's information), one may try to gather information from a statistical analysis of the past historic data combined with a model of how the price S_τ evolves as a function of time. The most typical model is the binomial model, which is a Wiener process. We assume that the time evolution of the price of the asset X is a stochastic process similar to a random walk. We divide time into intervals of size Δ_t, and we assume that in each time interval $\tau = n\Delta_t$, the variation in the asset price is

$$S_{n+1} = S_n u \text{ with probability } p \tag{7.56}$$
$$S_{n+1} = S_n d \text{ with probability } 1 - p \tag{7.57}$$

where $u > 1$ and $0 < d < 1$ are measures for historic data. It follows that for $\tau = n\Delta_t$, the probability that the spot price of the asset at expiration is $S_u u^i d^{n-i}$ is given by

$$\text{Prob}(S_\tau = S_u u^i d^{n-i}) = \binom{n}{i} p^i (1-p)^{n-i} \tag{7.58}$$

and therefore

$$C_{call} = e^{-r\tau} \frac{1}{n} \sum_{i=0}^{i \le n} \binom{n}{i} p^i (1-p)^{n-i} \max(S_u u^i d^{n-i} - A, 0) \tag{7.59}$$

and

$$C_{put} = e^{-r\tau} \frac{1}{n} \sum_{i=0}^{i \le n} \binom{n}{i} p^i (1-p)^{n-i} \max(A - S_u u^i d^{n-i}, 0) \tag{7.60}$$

The parameters of this model are u, d and p, and they must be determined from historical data. For example,

$$p = \frac{e^{r\Delta_t} - d}{u - d} \tag{7.61}$$

$$u = e^{\sigma\sqrt{\Delta_t}} \tag{7.62}$$

$$d = e^{-\sigma\sqrt{\Delta_t}} \tag{7.63}$$

where Δ_t is the length of the time interval, r is the risk-free rate, and σ is the volatility of the asset, that is, the standard deviation of the log returns.

Here is a Python code to simulate an asset price using a binomial tree:

```
def BinomialSimulation(S0,u,d,p,n):
    data=[]
    S=S0
    for i in xrange(n):
        data.append(S)
        if uniform()<p:
            S=u*S
        else:
            S=d*S
    return data
```

The function takes the present spot value, S_0, of the asset, the values of u, d and p, and the number of simulation steps and returns a list containing the simulated evolution of the stock price. Note that because of the exact formulas, eqs.(7.59) and (7.60), one does not need to perform a simulation unless the underlying asset is a stock that pays dividends or we want to include some other variable in the model.

This method works fine for European call options, but the method is not easy to generalize to other options, when its depends on the path of the asset (e.g., the asset is a stock that pays dividends). Moreover, to increase precision, one has to decrease Δ_t or redo the computation from the beginning.

The Monte Carlo method that we see next is slower in the simple cases but is more general and therefore more powerful.

7.6.2 Pricing European options: Monte Carlo

Here we adopt the Black–Scholes model assumptions. We assume that the time evolution of the price of the asset X is a stochastic process similar to a random walk [62]. We divide time into intervals of size Δ_t, and we assume that in each time interval $t = n\Delta_t$, the log return is a Gaussian random variable:

$$\log \frac{S_{n+1}}{S_n} = \text{gauss}(\mu\Delta_t, \sigma\sqrt{\Delta_t}) \tag{7.64}$$

There are three parameters in the preceding equation:

- Δ_t is the time step we use in our discretization. Δ_t is not a physical parameter; it has nothing to do with the asset. It has to do with the precision of our computation. Let's assume that $\Delta_t = 1$ day.

- μ is a drift term, and it represents the expected rate of return of the asset over a time scale of one year. It is usually set equal to the risk-free rate.

- σ is called volatility, and it represents the number of stochastic fluctuations of the asset over a time interval of one year.

Notice that this model is equivalent to the previous binomial model for large time intervals, in the same sense as the binomial distribution for large values of n approximates the Gaussian distribution. For large T, converge to the same result.

Notice how our assumption that log-return is Gaussian is different and not compatible with Markowitz's assumption of modern portfolio theory (the arithmetic return is Gaussian). In fact, log returns and arithmetic returns cannot both be Gaussian. It is therefore incorrect to optimize a portfolio using MPT when the portfolio includes options priced using Black–Scholes. The price of an individual asset cannot be negative, therefore its arithmetic return cannot be negative and it cannot be Gaussian. Conversely, a portfolio that includes both short and long positions (the holder is the buyer and seller of options) can have negative value. A change of sign in a portfolio is not compatible with the Gaussian log-return assumption.

If we are pricing a European call option, we are only interested in S_T and not in S_t for $0 < t < T$; therefore we can choose $\Delta_t = T$. In this case, we obtain

$$S_T = S_0 exp(r_T) \qquad (7.65)$$

and

$$p(r_T) \propto \exp\left(-\frac{(r_T - \mu T)^2}{2\sigma^2 T}\right) \qquad (7.66)$$

This allows us to write the following:

Listing 7.8: in file: `options.py`

```
from nlib import *

class EuropeanCallOptionPricer(MCEngine):
    def simulate_once(self):
        T = self.time_to_expiration
        S = self.spot_price
        R_T = random.gauss(self.mu*T, self.sigma*sqrt(T))
        S_T = S*exp(r_T)
        payoff = max(S_T-self.strike,0)
        return self.present_value(payoff)

    def present_value(self,payoff):
        daily_return = self.risk_free_rate/250
        return payoff*exp(-daily_return*self.time_to_expiration)

def main():
    pricer = EuropeanCallOptionPricer()
    # parameters of the underlying
    pricer.spot_price = 100 # dollars
    pricer.mu = 0.12/250 # daily drift term
    pricer.sigma = 0.30/sqrt(250) # daily variance
    # parameters of the option
    pricer.strike = 110 # dollars
    pricer.time_to_expiration = 90 # days
    # parameters of the market
    pricer.risk_free_rate = 0.05 # 5% annual return

    result = pricer.simulate_many(ap=0.01,rp=0.01) # precision: 1c or 1%
```

```
29      print result
30
31  main()
```

7.6.3   Pricing any option with Monte Carlo

An option is a contract, and one can write a contract with many different clauses. Each of them can be implemented into an algorithm. Yet we can group them into three different categories:

- Non-path-dependent: They depend on the price of the underlying asset at expiration but not on the intermediate prices of the asset (path).

- Weakly path-dependent: They depend on the price of the underlying asset and events that may happen to the price before expiration, but they do not depend on when the events exactly happen.

- Strongly path-dependent: They depend on the details of the time variation of price of the underlying asset before expiration.

Because non-path-dependent options do not depend on details, it is often possible to find approximate analytical formulas for pricing the option. For weakly path-dependent options, usually the binomial tree approach of the previous section is a preferable approach. The Monte Carlo approach applies to the general case, for example, that of strongly path-dependent options.

We will use our MCEngine to implement a generic option pricer.

First we need to recognize the following:

- The value of an option at expiration is defined by a payoff function $f(x)$ of the spot price of the asset at the expiration date. The fact that a call option has payoff $f(x) = \max(x - A, 0)$ is a convention that defined the European call option. A different type of option will have a different payoff function $f(x)$.

- The more accurately we model the underlying asset, the more accurate will be the computed value of the option. Some options are more sensitive than others to our modeling details.

Note one never model the option. One only model the underlying asset. The option payoff is given. We only choose the most efficient algorithm based on the model and the option:

Listing 7.9: in file: `options.py`

```
1  from nlib import *
2
3  class GenericOptionPricer(MCEngine):
4      def simulate_once(self):
5          S = self.spot_price
6          path = [S]
7          for t in range(self.time_to_expiration):
8              r = self.model(dt=1.0)
9              S = S*exp(r)
10             path.append(S)
11         return self.present_value(self.payoff(path))
12
13     def model(self,dt=1.0):
14         return random.gauss(self.mu*dt, self.sigma*sqrt(dt))
15
16     def present_value(self,payoff):
17         daily_return = self.risk_free_rate/250
18         return payoff*exp(-daily_return*self.time_to_expiration)
19
20     def payoff_european_call(self, path):
21         return max(path[-1]-self.strike,0)
22     def payoff_european_put(self, path):
23         return max(path[-1]-self.strike,0)
24     def payoff_exotic_call(self, path):
25         last_5_days = path[-5]
26         mean_last_5_days = sum(last_5_days)/len(last_5_days)
27         return max(mean_last_5_days-self.strike,0)
28
29 def main():
30     pricer = GenericOptionPricer()
31     # parameters of the underlying
32     pricer.spot_price = 100 # dollars
33     pricer.mu = 0.12/250 # daily drift term
34     pricer.sigma = 0.30/sqrt(250) # daily variance
35     # parameters of the option
36     pricer.strike = 110 # dollars
37     pricer.time_to_expiration = 90 # days
38     pricer.payoff = pricer.payoff_european_call
39     # parameters of the market
40     pricer.risk_free_rate = 0.05 # 5% annual return
41
42     result = pricer.simulate_many(ap=0.01,rp=0.01) # precision: 1c or 1%
```

```
43    print result
44
45  main()
```

This code allows us to price any option simply by changing the payoff function.

One can also change the model for the underlying using different assumptions. For example, a possible choice is that of including a model for market crashes, and on random days, separated by intervals given by the exponential distribution, assume a negative jump that follows the Pareto distribution (similar to the losses in our previous risk model). Of course, a change of the model requires a recalibration of the parameters.

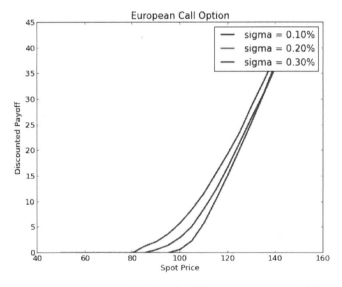

Figure 7.5: Price for a European call option for different spot prices and different values of σ.

7.7 Markov chain Monte Carlo (MCMC) and Metropolis

Until this point, all-out simulations were based on independent random variables. This means that we were able to generate each random number independently of the others because all the random variables were uncorrelated. There are cases when we have the following problem.

We have to generate $\mathbf{x} = x_0, x_1, ..., x_{n-1}$, where $x_0, x_1, ..., x_{n-1}$ are n correlated random variables whose probability mass function

$$p(\mathbf{x}) = p(x_0, x_1, ..., x_{n-1}) \qquad (7.67)$$

cannot be factored, as in $p(x_0, x_1, ..., x_{n-1}) = p(x_0)p(x_1)...p(x_{n-1})$. Consider for example the simple case of generating two random numbers x_0 and x_1 both in $[0, 1]$ with probability mass function $p(x_0, x_1) = 6(x_0 - x_1)^2$ (note that $\int_0^1 \int_0^1 6p(x_0, x_1)dx_0dx_1 = 1$, as it should be).

In the case where each of the x_i has a Gaussian distribution and the only dependence between x_i and x_j is their correlation, the solution was already examined in a previous section about the Cholesky algorithm. Here we examine the most general case.

The Metropolis algorithm provides a general and simpler solution to this problem. It is not always the most efficient, but more sophisticated algorithms are nothing but refinements and extensions of its simple idea.

Let's formulate the problem once more: we want to generate $\mathbf{x} = x_0, x_1, ..., x_{n-1}$ where $x_0, x_1, ..., x_{n-1}$ are n correlated random variables whose probability mass function is given by

$$p(\mathbf{x}) = p(x_0, x_1, ..., x_{n-1}) \qquad (7.68)$$

The procedure works as follows:

1 Start with a set of independent random numbers $\mathbf{x}^{(0)} = (x_0^{(0)}, x_1^{(0)}, ..., x_{n-1}^{(0)})$ in the domain.

2 Generate another set of independent random numbers $\mathbf{x}^{(i+1)} = (x_0^{(i+1)}, x_1^{(i+1)}, ..., x_{n-1}^{(i+1)})$ in the domain. This can be done by an arbitrary random function $Q(\mathbf{x}^{(i)})$. The only requirement for this function Q is that the probability of moving from a current point x to a new point y be the same as that of moving from a current point y to a new point x.

3 Generate a uniform random number z.

4 If $p(\mathbf{x}^{(i+1)})/p(\mathbf{x}^{(i)}) < z$, then $\mathbf{x}^{(i+1)} = \mathbf{x}^{(i)}$.

5 Go back to step 2.

The set of random numbers $\mathbf{x}^{(i)}$ generated in this way for large values of i will have a probability mass function given by $p(\mathbf{x})$.

Here is a possible implementation in Python:

```
def metropolis(p,q,x=None):
    while True:
        x_old=x
        x = q(x)
        if p(x)/p(x_old)<random.random()
            x=x_old
        yield x


def P(x):
    return 6.0*(x[0]-x[1])**2

def Q(x):
    return [random.random(), random.random()]

for i, x in enumerate(metropolis(P,Q)):
    print x
    if i==100: break
```

In this example, Q is the function that generates random points in the domain (in the example, $[0,1] \times [0,1]$), and P is an example probability $p(x) = 6(x_0 - x_1)^2$. Notice we used the Python yield function instead of return. This means the function is a generator and we can loop over its returned (yielded) values without having to generate all of them at once. They are generated as needed.

Notice that the Metropolis algorithm can generate (and will generate) repeated values. This is because the next random vector x is highly correlated with the previous vector. For this reason, it is often necessary to de-correlate metropolis values by skipping some of them:

```
def metropolis_decorrelate(p,q,x=None,ds=100):
    k = 0
    for x in metropolis(p,q,x):
        k += 1
        if k % ds == ds-1:
            yield x
```

The value of ds must be fixed empirically. The value of ds which is large enough to make the next vector independent from the previous one is called *decorrelation length*. This generator works as the previous one. For example:

```
1  for i, x in enumerate(metropolis_decorrelate(P,Q)):
2      print x
3      if i==100: break
```

7.7.1 The Ising model

A typical example of application of the Metropolis is the Ising model. This model describes a spin system, for example, a ferromagnet. A spin system consists of a regular crystalline structure, and each vertex is an atom. Each atom is a small magnet, and its magnetic orientation can be +1 or −1. Each atom interacts with the external magnetic field and with the magnetic field of its six neighbors (think about the six faces of a cube). We use the index i to label an atom and s_i its spin.

The entire system has a total energy given by

$$E(s) = -\sum_i s_i h - \sum_{ij|dist_{ij}=1} s_i s_j \tag{7.69}$$

where h is the external magnetic field, the first sum is over all spin sites, and the second is about all couples of next neighbor sites. In the absence of spin-spin interaction, only the first term contributes, and the energy is lower when the direction of the s_i (their sign) is the same as h. In absence of h, only the second term contributes. The contribution to each couple of spins is positive if their sign is the opposite, and negative otherwise.

In the absence of external forces, each system evolves toward the state of minimum energy, and therefore, for a spin system, each spin tends to align itself in the same direction as its neighbors and in the same direction as the external field h.

Things change when we turn on heat. Feeding energy to the system makes the atoms vibrate and the spins randomly flip. The higher the temperature, the more they randomly flip.

The probability of finding the system in a given state s at a given temperature T is given by the Boltzmann distribution:

$$p(s) = \exp\left(-\frac{E(s)}{KT}\right) \qquad (7.70)$$

where K is the Boltzmann constant.

We can now use the Metropolis algorithm to generate possible states of the system s compatible with a given temperature T and measure the effects on the average magnetization (the average spin) as a function of T and possibly an external field h.

Also notice that in the case of the Boltzmann distribution,

$$\frac{p(s')}{p(s)} = exp\left(\frac{E(s) - E(s')}{KT}\right) \qquad (7.71)$$

only depends on the change in energy. The Metropolis algorithm gives us the freedom to choose a function Q that changes the state of the system and depends on the current state. We can choose such a function so that we only try to flip one spin at a time. In this case, the P algorithm simplifies because we no longer need to compute the total energy of the system at each iteration, but only the variation of energy due to the flipping of that one spin.

Here is the code for a three-dimensional spin system:

Listing 7.10: in file: `ising.py`

```
import random
import math
from nlib import Canvas, mean, sd

class Ising:
    def __init__(self, n):
        self.n = n
        self.s = [[[1 for x in xrange(n)] for y in xrange(n)]
                  for z in xrange(n)]
        self.magnetization = n**3
```

```
12    def __getitem__(self,point):
13        n = self.n
14        x,y,z = point
15        return self.s[(x+n)%n][(y+n)%n][(z+n)%n]
16
17    def __setitem__(self,point,value):
18        n = self.n
19        x,y,z = point
20        self.s[(x+n)%n][(y+n)%n][(z+n)%n] = value
21
22    def step(self,t,h):
23        n = self.n
24        x,y,z = random.randint(0,n-1),random.randint(0,n-1),random.randint(0,n
              -1)
25        neighbors = [(x-1,y,z),(x+1,y,z),(x,y-1,z),(x,y+1,z),(x,y,z-1),(x,y,z+1)
              ]
26        dE = -2.0*self[x,y,z]*(h+sum(self[xn,yn,zn] for xn,yn,zn in neighbors))
27        if dE > t*math.log(random.random()):
28            self[x,y,z] = -self[x,y,z]
29            self.magnetization += 2*self[x,y,z]
30        return self.magnetization
31
32 def simulate(steps=100):
33     ising = Ising(n=10)
34     data = {}
35     for h in range(0,11): # external magnetic field
36         data[h] = []
37         for t in range(1,11): # temperature, in units of K
38             m = [ising.step(t=t,h=h) for k in range(steps)]
39             mu = mean(m) # average magnetization
40             sigma = sd(m)
41             data[h].append((t,mu,sigma))
42     return data
43
44 def main(name='ising.png'):
45     data = simulate(steps = 10000)
46     canvas = Canvas(xlab='temperature', ylab='magnetization')
47     for h in data:
48         color = '#%.2x0000' % (h*25)
49         canvas.errorbar(data[h]).plot(data[h],color=color)
50     canvas.save(name)
51
52 main()
```

Fig. 7.7.1 shows how the spins tend to align in the direction of the external
magnetic field, but the larger the temperature (left to right), the more
random they are, and the average magnetization tends to zero. The higher
the external magnetic field (bottom to top curves), the longer it takes for

the transition from order (aligned spins) to chaos (random spins).

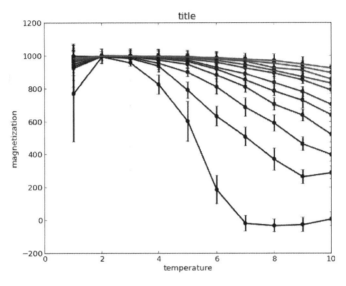

Figure 7.6: Average magnetization as a function of the temperature for a spin system.

Fig. 7.7.1 shows the two-dimensional section of some random three-dimensional states for different values of the temperature. One can clearly see that the lower the temperature, the more the spins are aligned, and the higher the temperature, the more random they are.

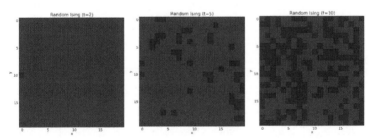

Figure 7.7: Random Ising states (2D section of 3D) for different temperatures.

7.8   Simulated annealing

Simulated annealing is an application of Monte Carlo to solve optimization problems. It is best understood within the context of the Ising model. When the temperature is lowered, the system tends toward the state of minimum energy. At high temperature, the system fluctuates randomly and moves in the space of all possible states. This behavior is not specific to the Ising model. Hence, for any system for which we can define an energy, we can find its minimum energy state, by starting in a random state and slowly lowering the temperature as we evolve the simulation. The system will find a minimum. There may be more than one minimum, and one may need to repeat the procedure multiple times from different initial random states and compare the solutions. This process takes the name of annealing in analogy with the industrial process for removing impurities from metals: heat, cool slowly, repeat.

We can apply this process to any system for which we want to minimize a function $f(x)$ of multiple variables. We just have to think of x as the state s and of f as the energy E. This analogy is purely semantic because the quantity we want to minimize is not necessarily an energy in the physical sense.

Simulated annealing does not assume the function is differentiable or continuous in its variables.

7.8.1   Protein folding

In the following we apply simulated annealing to the problem of folding of a protein. A protein is a sequence of amino-acids. It is normally unfolded, and amino-acids are on a line. When placed in water, it folds. This is because some amino-acids are hydrophobic (repel water) and some are hydrophilic (like contact with water), therefore the protein tries to acquire a three-dimensional shape that minimizes the surface of hydrophobic amino-acids in contact with water [63]. This is represented graphically in fig. 7.8.1.

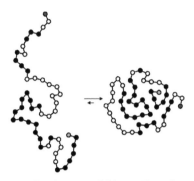

Figure 7.8: Schematic example of protein folding. The white circles are hydrophilic amino-acids. The black ones are hydrophobic.

Here we assume only two types of amino-acids (H for hydrophobic and P for hydrophilic), and we assume each amino acid is a cube, that all cubes have the same size, and that each two consecutive amino-acids are connected at a face. These assumptions greatly simplify the problem because they limit the possible solid angles to six possible values (0: up, 1: down, 2: right, 3: left, 4: front, 5: back). Our goal is arranging the cubes to minimize the number of faces of hydrophobic cubes that are exposed to water:

Listing 7.11: in file: `folding.py`

```python
import random
import math
import copy
from nlib import *

class Protein:

    moves = {0:(lambda x,y,z: (x+1,y,z)),
             1:(lambda x,y,z: (x-1,y,z)),
             2:(lambda x,y,z: (x,y+1,z)),
             3:(lambda x,y,z: (x,y-1,z)),
             4:(lambda x,y,z: (x,y,z+1)),
             5:(lambda x,y,z: (x,y,z-1))}

    def __init__(self, aminoacids):
        self.aminoacids = aminoacids
        self.angles = [0]*(len(aminoacids)-1)
        self.folding = self.compute_folding(self.angles)
```

```
19      self.energy = self.compute_energy(self.folding)
20
21  def compute_folding(self,angles):
22      folding = {}
23      x,y,z = 0,0,0
24      k = 0
25      folding[x,y,z] = self.aminoacids[k]
26      for angle in angles:
27          k += 1
28          xn,yn,zn = self.moves[angle](x,y,z)
29          if (xn,yn,zn) in folding: return None # impossible folding
30          folding[xn,yn,zn] = self.aminoacids[k]
31          x,y,z = xn,yn,zn
32      return folding
33
34  def compute_energy(self,folding):
35      E = 0
36      for x,y,z in folding:
37          aminoacid = folding[x,y,z]
38          if aminoacid == 'H':
39              for face in range(6):
40                  if not self.moves[face](x,y,z) in folding:
41                      E = E + 1
42      return E
43
44  def fold(self,t):
45      while True:
46          new_angles = copy.copy(self.angles)
47          n = random.randint(1,len(self.aminoacids)-2)
48          new_angles[n] = random.randint(0,5)
49          new_folding = self.compute_folding(new_angles)
50          if new_folding: break # found a valid folding
51      new_energy = self.compute_energy(new_folding)
52      if (self.energy-new_energy) > t*math.log(random.random()):
53          self.angles = new_angles
54          self.folding = new_folding
55          self.energy = new_energy
56      return self.energy
57
58  def main():
59      aminoacids = ''.join(random.choice('HP') for k in range(20))
60      protein = Protein(aminoacids)
61      t = 10.0
62      while t>1e-5:
63          protein.fold(t = t)
64          print protein.energy, protein.angles
65          t = t*0.99 # cool
66
67  main()
```

The moves dictionary is a dictionary of functions. For each solid angle (0–5), moves[angle] is a function that maps x,y,z, the coordinates of an amino acid, to the coordinates of the cube at that solid angle.

The annealing procedure is performed in the main function. The fold procedure is the same step as the Metropolis step. The purpose of the while loop in the fold function is to find a valid fold for the accept–reject step. Some folds are invalid because they are not physical and would require two amino-acids to occupy the same portion of space. When this happens, the compute_folding method returns None, indicating that one must try a different folding.

8

Parallel Algorithms

Consider a program that performs the following computation:

```
1  y = f(x)
2  z = g(x)
```

In this example, the function $g(x)$ does not depend on the result of the function $f(x)$, and therefore the two functions could be computed independently and in parallel.

Often large problems can be divided into smaller computational problems, which can be solved concurrently ("in parallel") using different processing units (CPUs, cores). This is called *parallel computing*. Algorithms designed to work in parallel are called *parallel algorithms*.

In this chapter, we will refer to a processing unit as a node and to the code running on a node as a process. A parallel program consists of many processes running on as many nodes. It is possible for multiple processes to run on one and the same computing unit (node) because of the multitasking capabilities of modern CPUs, but that is not true parallel computing. We will use an emulator, Psim, which does exactly that.

Programs can be parallelized at many levels: bit level, instruction level, data, and task parallelism. Bit-level parallelism is usually implemented in hardware. Instruction-level parallelism is also implemented in hardware in modern multi-pipeline CPUs. Data parallelism is usually referred to as

SIMD. Task parallelism is also referred to as MIMD.

Historically, parallelism was found in applications in high-performance computing, but today it is employed in many devices, including common cell phones. The reason is heat dissipation. It is getting harder and harder to improve speed by increasing CPU frequency because there is a physical limit to how much we can cool the CPU. So the recent trend is keeping frequency constant and increasing the number of processing units on the same chip.

Parallel architectures are classified according to the level at which the hardware supports parallelism, with multicore and multiprocessor computers having multiple processing elements within a single machine, while clusters, MPPs, and grids use multiple computers to work on the same task. Specialized parallel computer architectures are sometimes used alongside traditional processors for accelerating specific tasks.

Optimizing an algorithm to run on a parallel architecture is not an easy task. Details depend on the type of parallelism and details of the architecture.

In this chapter, we will learn how to classify architectures, compute running times of parallel algorithms, and measure their performance and scaling.

We will learn how to write parallel programs using standard programming patterns, and we will use them as building blocks for more complex algorithms.

For some parts of this chapter, we will use a simulator called PSim, which is written in Python. Its performances will only scale on multicore machines, but it will allow us to emulate various network topologies.

8.1 Parallel architectures

8.1.1   Flynn taxonomy

Parallel computer architecture classifications are known as Flynn's taxonomy [64] and are due to the work of Michael J. Flynn in 1966.

Flynn identified the following architectures:

- **Single instruction, single data stream (SISD)**

  A sequential computer that exploits no parallelism in either the instruction or data streams. A single control unit (CU) fetches a single instruction stream (IS) from memory. The CU then generates appropriate control signals to direct single processing elements (PE) to operate on a single data stream (DS), for example, one operation at a time.

  Examples of SISD architecture are the traditional uniprocessor machines like a PC (currently manufactured PCs have multiple processors) or old mainframes.

- **Single instruction, multiple data streams (SIMD)** A computer that exploits multiple data streams against a single instruction stream to perform operations that may be naturally parallelized (e.g., an array processor or GPU).

- **Multiple instruction, single data stream (MISD)**

  Multiple instructions operate on a single data stream. This is an uncommon architecture that is generally used for fault tolerance. Heterogeneous systems operate on the same data stream and must agree on the result. Examples include the now retired Space Shuttle flight control computer.

- **Multiple instruction, multiple data streams (MIMD)** Multiple autonomous processors simultaneously executing different instructions on different data. Distributed systems are generally recognized to be MIMD architectures, either exploiting a single shared memory space (using threads) or a distributed memory space (using a message-passing protocol such as MPI).

MIMD can be further subdivided into the following:

- **Single program, multiple data (SPMD)** Multiple autonomous processors simultaneously executing the same program but at independent points not synchronously (as in the SIMD case). SPMD is the most common style of parallel programming.

- **Multiple program, multiple data (MPMD)** Multiple autonomous processors simultaneously operating at least two independent programs. Typically such systems pick one node to be the "host" ("the explicit host/node programming model") or "manager" (the "manager–worker" strategy), which runs one program that farms out data to all the other nodes, which all run a second program. Those other nodes then return their results directly to the manager. The Map-Reduce pattern also falls under this category.

An embarrassingly parallel workload (or embarrassingly parallel problem) is one for which little or no effort is required to separate the problem into a number of parallel tasks. This is often the case where there exists no dependency (or communication) between those parallel tasks.

The manager–worker node strategy, when workers do not need to communicate with each other, is an example of an "embarrassingly parallel" problem.

8.1.2 Network topologies

In the MIMD case, multiple copies of the same problem run concurrently (on different data subsets and branching differently, thus performing different instructions) on different processing units, and they exchange information using a network. How fast they can communicate depends on the network characteristics identified by the network topology and the latency and bandwidth of the individual links of the network.

Normally we classify network topologies based on the following taxonomy:

- **Completely connected:** Each node is connected by a directed link to each other node.

- **Bus topology:** All nodes are connected to the same single cable. Each computer can therefore communicate with each other computer using one and the same bus cable. The limitation of this approach is that the communication bandwidth is limited by the bandwidth of the cable. Most bus networks only allow two machines to communicate with each other at one time (with the exception of one too many broadcast messages). While two machines communicate, the others are stuck waiting. The bus topology is the most inexpensive but also slow and constitutes a single point of failure.

- **Switch topology (star topology):** In local area networks with a switch topology, each computer is connected via a direct link to a central device, usually a switch, and it resembles a star. Two computers can communicate using two links (to the switch and from the switch). The central point of failure is the switch. The switch is usually intelligent and can reroute the messages from any computer to any other computer. If the switch has sufficient bandwidth, it can allow multiple computers to talk to each other at the same time. For example, for a 10 Gbit/s links and an 80 Gbit/s switch, eight computers can talk to each other (in pairs) at the same time.

- **Mesh topology:** In a mesh topology, computers are assembled into an array (1D, 2D, etc.), and each computer is connected via a direct link to the computers immediately close (left, right, above, below, etc.). Next neighbor communication is very fast because it involves a single link and therefore low latency. For two computers not physically close to communicate, it is necessary to reroute messages. The latency is proportional to the distance in links between the computers. Some meshes do not support this kind of rerouting because the extra logic, even if unused, may be cause for extra latency. Meshes are ideal for solving numerical problems such as solving differential equations because they can be naturally mapped into this kind of topology.

- **Torus topology:** Very similar to a mesh topology (1D, 2D, 3D, etc.), except that the network wraps around the edges. For example, in one dimension node, i is connected to $(i + 1)\%p$, where p is the total number of nodes. A one-dimensional torus is called a *ring network*.

- **Tree network:** The tree topology looks like a tree where the computer may be associated with every tree node or every leaf only. The tree links are the communication link. For a binary tree, each computer only talks to its parent and its two children nodes. The root node is special because it has no parent node.

  Tree networks are ideal for global operations such as broadcasting and for sharing IO devices such as disks. If the IO device is connected to the root node, every other computer can communicate with it using only $\log p$ links (where p is the number of computers connected). Moreover, each subset of a tree network is also a tree network. This makes it easy to distribute subtasks to different subsets of the same architecture.

- **Hypercube:** This network assumes 2^d nodes, and each node corresponds to a vertex of a hypercube. Nodes are connected by direct links, which correspond to the edges of the hypercube. Its importance is more academic than practical, although some ideas from hypercube networks are implemented in some algorithms.

If we identify each node on the network with a unique integer number called its rank, we write explicit code to determine if two nodes i and j are connected for each network topology:

Listing 8.1: in file: `psim.py`

```
 1  import os, string, pickle, time, math
 2
 3  def BUS(i,j):
 4      return True
 5
 6  def SWITCH(i,j):
 7      return True
 8
 9  def MESH1(p):
10      return lambda i,j,p=p: (i-j)**2==1
11
12  def TORUS1(p):
13      return lambda i,j,p=p: (i-j+p)%p==1 or (j-i+p)%p==1
14
15  def MESH2(p):
16      q=int(math.sqrt(p)+0.1)
17      return lambda i,j,q=q: ((i%q-j%q)**2,(i/q-j/q)**2) in [(1,0),(0,1)]
18
```

```
19  def TORUS2(p):
20      q=int(math.sqrt(p)+0.1)
21      return lambda i,j,q=q: ((i%q-j%q+q)%q,(i/q-j/q+q)%q) in [(0,1),(1,0)] or \
22                             ((j%q-i%q+q)%q,(j/q-i/q+q)%q) in [(0,1),(1,0)]
23  def TREE(i,j):
24      return i==int((j-1)/2) or j==int((i-1)/2)
```

8.1.3   Network characteristics

- **Number of links**

- **Diameter:** The max distance between any two nodes measured as a minimum number of links connecting them. Smaller diameter means smaller latency. The diameter is proportional to the maximum time it takes for a message go from one node to another.

- **Bisection width:** The minimum number of links one has to cut to turn the network into two disjoint networks. Higher bisection width means higher reliability of the network.

- **Arc connectivity:** The number of different paths (non-overlapping and of minimal length) connecting any two nodes. Higher connectivity means higher bandwidth and higher reliability.

Here are values of this parameter for each type of network:

Network	Links	Diameter	Width
completely connected	$p(p-1)/2$	1	$p-1$
switch	p	2	1
1D mesh	$p-1$	$p-1$	1
nD mesh	$n(p^{\frac{1}{n}}-1)p^{\frac{n-1}{n}}$	n	$p^{\frac{2}{3}}$
1D torus	p	$\frac{p}{2}$	2
nD torus	np	$\frac{n}{2}p^{\frac{1}{n}}$	$2n$
hypercube	$\frac{p}{2}\log_2 p$	$\log_2 p$	$log_2 p$
tree	$p-1$	$\log_2 p$	1

Most actual supercomputers implement a variety of taxonomies and topologies simultaneously. A modern supercomputer has many nodes, each node has many CPUs, each CPU has many cores, and each core im-

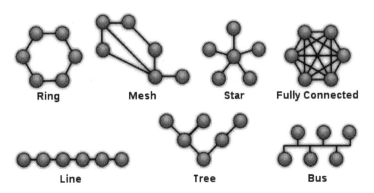

Figure 8.1: Examples of network topologies.

plements SIMD instructions. Each core has it own cache, each CPU has its own cache, and each node has its own memory shared by all threads running on that one node. Nodes communicate with each other using multiple networks (typically a multidimensional mesh for point-to-point communications and a tree network for global communication and general disk IO).

This makes writing parallel programs very difficult. Parallel programs must be optimized for each specific architecture.

8.2 Parallel metrics

8.2.1 Latency and bandwidth

The time it takes for a message of size m (in bytes) over a wire can be broken into two components: a fixed overall time that does not depend on the size of the message, called *latency* (and indicated with t_s), and a time proportional to the message size, called *inverse bandwidth* (and indicated with t_w).

Think of a pipe of length L and section s, and you want to pump m liters of water through the pipe at velocity v. From the moment you start pumping, it takes L/v seconds before the water starts arriving at the other end of the pipe. From that moment, it will take m/sv for all the water to arrive at its destination. In this analogy, L/v is the latency t_s, sv is the bandwidth, and $t_w = 1/sv$.

The total time to send the message (or the water) is

$$T(m) = t_s + t_w m \qquad (8.1)$$

From now on, we will use $T_1(n)$ to refer to the nonparallel running time of an algorithm as a function of its input m. We will use $T_p(n)$ to refer to its running time with p parallel processes.

As a practical case, in the following example, we consider a generic algorithm with the following parallel and nonparallel running times:

$$
\begin{aligned}
T_1(n) &= t_a n^2 & (8.2) \\
T_p(n) &= t_a n^2/p + 2p(t_s + t_w n/p) & (8.3)
\end{aligned}
$$

These formulas may come from example from the problem of multiplying a matrix times a vector.

Here t_a is the time to perform one elementary instruction; t_s and t_w are the latency and inverse bandwidth. The first term of T_p is nothing but T_1/p, while the second term is an overhead due to communications.

Typically $t_s \gg t_w \gg t_a$. In the following plots, we will always assume $t_a = 1$, $t_s = 0.2$, and $t_w = 0.1$. With these assumptions, fig. 8.2.1 shows how T_p changes with input size and number of parallel processes. Notice that while for small p, T_p decreases $\propto 1/p$, for large p, the communication overhead dominates over computation. This overhead is our example and is dominated by the latency contribution, which grows with p.

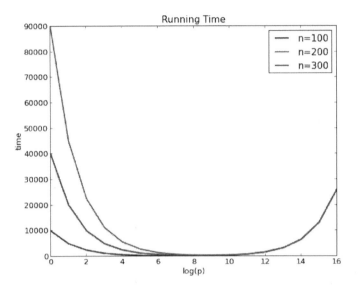

Figure 8.2: T_p as a function of input size n and number of processes p.

8.2.2 Speedup

The *speedup* is defined as

$$S_p(n) = \frac{T_1(n)}{T_p(n)} \tag{8.4}$$

where T_1 is the time it takes to run the algorithm on an input of size n on one processing unit (e.g., node), and T_p is the time it takes to run the same algorithm on the same input using p nodes in parallel. Fig. 8.2.2 shows an example of speedup. When communication overhead dominates, speedup decreases.

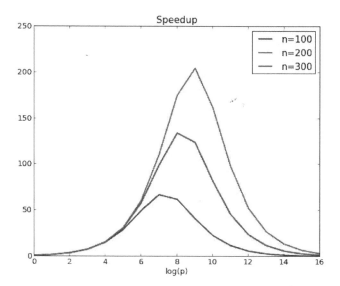

Figure 8.3: S_p as a function of input size n and number of processes p.

8.2.3   Efficiency

The *efficiency* is defined as

$$E_p(n) = \frac{S_p(n)}{p} = \frac{T_1(n)}{pT_p(n)} \tag{8.5}$$

Notice that in case of perfect parallelization (impossible), $T_p = T_1/p$, and therefore $E_p(n) = 1$. Fig. 8.2.3 shows an example of efficiency. When communication overhead dominates, efficiency drops. Notice efficiency is always less than 1. We do not write parallel algorithms because they are more efficient. They are always less efficient and more costly than the nonparallel ones. We do it because we want the result sooner, and there is an economic value in it.

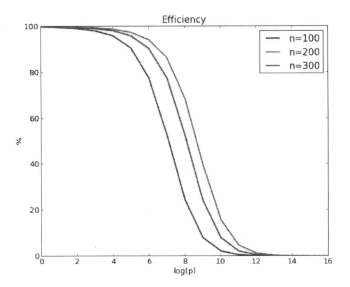

Figure 8.4: E_p as a function of input size n and number of processes p.

8.2.4   Isoefficiency

Given a value of efficiency that we choose as target, E, and a given number of nodes, p, we ask what is the maximum size of a problem that we can solve. The answer is found by solving n the following equation:

$$E_p(n) = E \qquad (8.6)$$

For example T_p, we obtain

$$E_p = \frac{1}{1 + 2p^2(t_s + t_w n/p)/(n^2 t_a)} = E \qquad (8.7)$$

which solved in n yields

$$n \simeq 2\frac{t_w}{t_a}\frac{E}{1-E}p \qquad (8.8)$$

Isoefficiency curves for different values of E are shown in fig. 8.2.4. For our example problem, n is proportional to p. In general, this is not true, but n is monotonic in p.

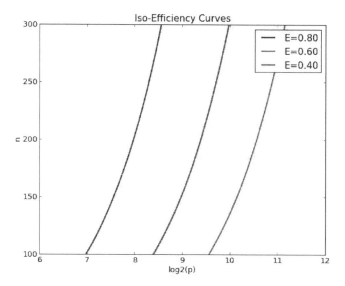

Figure 8.5: Isoefficiency curves for different values of the target efficiency.

8.2.5   Cost

The cost of a computation is equal to the time it takes to run on each node, multiplied by the number of nodes involved in the computation:

$$C_p(n) = pT_p(n) \qquad (8.9)$$

Notice that in general

$$\frac{\mathrm{d}C_p(n)}{\mathrm{d}p} = \alpha T_1(n) > 0 \qquad (8.10)$$

This means that for a fixed problem size n, the more an algorithm is parallelized, the more it costs to run it (because it gets less and less efficient).

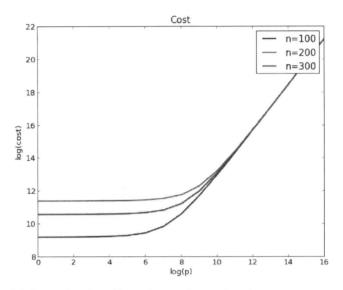

Figure 8.6: C_p as a function of input size n and a number of processes p.

8.2.6 Cost optimality

With the preceding disclaimer, we define cost optimality as the choice of p (as a function of n), which makes the cost scale proportional to $T_1(n)$:

$$pT_p(n) \propto T_1(n) \qquad (8.11)$$

Or in other words, looking for the $p(n)$ such that

$$\lim_{n \to \infty} p(n)T_{p(n)}(n)/T_1(n) = const. \neq 0 \qquad (8.12)$$

8.2.7 Admahl's law

Consider an algorithm that can be parallelized, but one faction α of its total sequential running time αT_1 cannot be parallelized. That means that

$T_p = \alpha T_1 + (1 - \alpha)T_1/p$, and this yields [65]

$$S_p = \frac{1}{\alpha + (1 - \alpha)/p} < \frac{1}{\alpha} \qquad (8.13)$$

Therefore the speedup is theoretically limited.

8.3   Message passing

Consider the following Python program:

```
def f():
    import os
    if os.fork(): print True
    else: print False
f()
```

The output of the current program is

```
True
False
```

The function fork creates a copy of the current process (a child). The parent process returns the ID of the child process, and the child process returns 0. Therefore the if condition is both true and false, just on different processes.

We have created a Python module called psim, and its source code is listed here; psim forks the parallel processes, creates sockets connecting them, and provides API for communications. An example of psim usage will be given later.

Listing 8.2: in file: `psim.py`

```
class PSim(object):
    def log(self,message):
        """

        logs the message into self._logfile
        """

        if self.logfile!=None:
            self.logfile.write(message)

    def __init__(self,p,topology=SWITCH,logfilename=None):
        """
```

```
        forks p-1 processes and creates p*p
        """
        self.logfile = logfilename and open(logfile,'w')
        self.topology = topology
        self.log("START: creating %i parallel processes\n" % p)
        self.nprocs = p
        self.pipes = {}
        for i in xrange(p):
            for j in xrange(p):
                self.pipes[i,j] = os.pipe()
        self.rank = 0
        for i in xrange(1,p):
            if not os.fork():
                self.rank = i
                break
        self.log("START: done.\n")

    def send(self,j,data):
        """
        sends data to process #j
        """
        if not self.topology(self.rank,j):
            raise RuntimeError('topology violation')
        self._send(j,data)

    def _send(self,j,data):
        """
        sends data to process #j ignoring topology
        """
        if j<0 or j>=self.nprocs:
            self.log("process %i: send(%i,...) failed!\n" % (self.rank,j))
            raise Exception
        self.log("process %i: send(%i,%s) starting...\n" % \
                (self.rank,j,repr(data)))
        s = pickle.dumps(data)
        os.write(self.pipes[self.rank,j][1], string.zfill(str(len(s)),10))
        os.write(self.pipes[self.rank,j][1], s)
        self.log("process %i: send(%i,%s) success.\n" % \
                (self.rank,j,repr(data)))

    def recv(self,j):
        """
        returns the data received from process #j
        """
        if not self.topology(self.rank,j):
            raise RuntimeError('topology violation')
        return self._recv(j)

    def _recv(self,j):
```

```
60      """
61      returns the data received from process #j ignoring topology
62      """
63      if j<0 or j>=self.nprocs:
64          self.log("process %i: recv(%i) failed!\n" % (self.rank,j))
65          raise RuntimeError
66      self.log("process %i: recv(%i) starting...\n" % (self.rank,j))
67      try:
68          size=int(os.read(self.pipes[j,self.rank][0],10))
69          s=os.read(self.pipes[j,self.rank][0],size)
70      except Exception, e:
71          self.log("process %i: COMMUNICATION ERROR!!!\n" % (self.rank))
72          raise e
73      data=pickle.loads(s)
74      self.log("process %i: recv(%i) done.\n" % (self.rank,j))
75      return data
```

An instance of the class PSim is an object that can be used to determine the total number of parallel processes, the rank of each running process, to send messages to other processes, and to receive messages from them. It is usually called a *communicator*; send and recv represent the simplest type of communication pattern, point-to-point communication.

A PSim program starts by importing and creating an instance of the PSim class. The constructor takes two arguments, the number of parallel processes you want and the network topology you want to emulate. Before returning the PSim instance, the constructor makes $p - 1$ copies of the running process and creates sockets connecting each two of them. Here is a simple example in which we make two parallel processes and send a message from process 0 to process 1:

```
1  from psim import *
2
3  comm = PSim(2,SWITCH)
4  if comm.rank == 0:
5      comm.send(1, "Hello World")
6  elif comm.rank == 1:
7      message = comm.recv(0)
8      print message
```

Here is a more complex example that creates $p = 10$ parallel processes, and node 0 sends a message to each one of them:

```
1  from psim import *
2
3  p = 10
```

```
4
5   comm = PSim(p,SWITCH)
6   if comm.rank == 0:
7       for other in range(1,p):
8           comm.send(other, "Hello %s" % p)
9   else:
10      message = comm.recv(0)
11      print message
```

Following is a more complex example that implements a parallel scalar product. The process with rank 0 makes up two vectors and distributes pieces of them to the other processes. Each process computes a part of the scalar product. Of course, the scalar product runs in linear time, and it is very inefficient to parallelize it, yet we do it for didactic purposes.

Listing 8.3: in file: `psim_scalar.py`

```
1   import random
2   from psim import PSim
3
4   def scalar_product_test1(n,p):
5       comm = PSim(p)
6       h = n/p
7       if comm.rank==0:
8           a = [random.random() for i in xrange(n)]
9           b = [random.random() for i in xrange(n)]
10          for k in xrange(1,p):
11              comm.send(k, a[k*h:k*h+h])
12              comm.send(k, b[k*h:k*h+h])
13      else:
14          a = comm.recv(0)
15          b = comm.recv(0)
16      scalar = sum(a[i]*b[i] for i in xrange(h))
17      if comm.rank == 0:
18          for k in xrange(1,p):
19              scalar += comm.recv(k)
20          print scalar
21      else:
22          comm.send(0,scalar)
23
24  scalar_product_test(10,2)
```

Most parallel algorithms follow a similar pattern. One process has access to IO. That process reads and scatters the data. The other processes perform their part of the computation; the results are reduced (aggregated) and sent back to the root process. This pattern may be repeated by multiple functions, perhaps in loops. Different functions may handle different data structure and may have different communication patterns. The one thing that must be constant throughout the run is the number of processes because one wants to pair each process with one computing unit.

In the following, we implement a parallel version of the mergesort. At each step, the code splits the problem into two smaller problems. Half of the problem is solved by the process that performed the split and assigns the other half to an existing free process. When there are no more free processes, it reverts to the nonparallel mergesort step. The merge function here is the same as the nonparallel mergesort of chapter 3.

Listing 8.4: in file: psim_mergesort.py

```
import random
from psim import PSim

def mergesort(A, x=0, z=None):
    if z is None: z = len(A)
    if x<z-1:
        y = int((x+z)/2)
        mergesort(A,x,y)
        mergesort(A,y,z)
        merge(A,x,y,z)

def merge(A,x,y,z):
    B,i,j = [],x,y
    while True:
        if A[i]<=A[j]:
            B.append(A[i])
            i=i+1
        else:
            B.append(A[j])
            j=j+1
        if i==y:
            while j<z:
                B.append(A[j])
                j=j+1
            break
        if j==z:
            while i<y:
                B.append(A[i])
                i=i+1
            break
    A[x:z]=B

def mergesort_test(n,p):
```

```
34    comm = PSim(p)
35    if comm.rank==0:
36        data = [random.random() for i in xrange(n)]
37        comm.send(1, data[n/2:])
38        mergesort(data,0,n/2)
39        data[n/2:] = comm.recv(1)
40        merge(data,0,n/2,n)
41        print data
42    else:
43        data = comm.recv(0)
44        mergesort(data)
45        comm.send(0,data)
46
47 mergesort_test(20,2)
```

More interesting patterns are global communication patterns implemented on top of send and recv. Subsequently, we discuss the most common: broadcast, scatter, collect, and reduce. Our implementation is not the most efficient, but it is the simplest. In principle, there should be a different implementation for each type of network topology to take advantage of its features.

8.3.1   Broadcast

The simplest type of broadcast is the one-2-all, which consists of one process (source) sending a message (value) to every other process. A more complex broadcast is when each process broadcasts a message simultaneously and each node receives the list of values ordered by the rank of the sender:

Listing 8.5: in file: `psim.py`

```
1    def one2all_broadcast(self, source, value):
2        self.log("process %i: BEGIN one2all_broadcast(%i,%s)\n" % \
3                (self.rank,source, repr(value)))
4        if self.rank==source:
5            for i in xrange(0, self.nprocs):
6                if i!=source:
7                    self._send(i,value)
8        else:
9            value=self._recv(source)
10        self.log("process %i: END one2all_broadcast(%i,%s)\n" % \
11                (self.rank,source, repr(value)))
12        return value
```

```
13
14    def all2all_broadcast(self, value):
15        self.log("process %i: BEGIN all2all_broadcast(%s)\n" % \
16            (self.rank, repr(value)))
17        vector=self.all2one_collect(0,value)
18        vector=self.one2all_broadcast(0,vector)
19        self.log("process %i: END all2all_broadcast(%s)\n" % \
20            (self.rank, repr(value)))
21        return vector
```

We have implemented the all-to-all broadcast using a trick. We send collected all values to node with rank 0 (via a function collect), and then we did a one-to-all broadcast of the entire list from node 0. In general, the implementation depends on the topology of the available network.

Here is an example of an application of broadcasting:

```
1  from psim import *
2
3  p = 10
4
5  comm = PSim(p,SWITCH)
6  message = "Hello World" if comm.rank==0 else None
7  message = comm.one2all_broadcast(0, message)
8  print message
```

Notice how before the broadcast, only the process with rank 0 has knowledge of the message. After broadcast, all nodes are aware of it. Also notice that one2all_broadcast is a global communication function, and all processes must call it. Its first argument is the rank of the broadcasting process (0), while the second argument is the message to be broadcasted (only the value from node 0 is actually used).

8.3.2   Scatter and collect

The all-to-one collect pattern works as follows. Every process sends a value to process destination, which receives the values in a list ordered according to the rank of the senders:

Listing 8.6: in file: psim.py

```
1    def one2all_scatter(self,source,data):
2        self.log('process %i: BEGIN all2one_scatter(%i,%s)\n' % \
3            (self.rank,source,repr(data)))
```

```
4      if self.rank==source:
5          h, remainder = divmod(len(data),self.nprocs)
6          if remainder: h+=1
7          for i in xrange(self.nprocs):
8              self._send(i,data[i*h:i*h+h])
9      vector = self._recv(source)
10     self.log('process %i: END all2one_scatter(%i,%s)\n' % \
11             (self.rank,source,repr(data)))
12     return vector
13
14 def all2one_collect(self,destination,data):
15     self.log("process %i: BEGIN all2one_collect(%i,%s)\n" % \
16             (self.rank,destination,repr(data)))
17     self._send(destination,data)
18     if self.rank==destination:
19         vector = [self._recv(i) for i in xrange(self.nprocs)]
20     else:
21         vector = []
22     self.log("process %i: END all2one_collect(%i,%s)\n" % \
23             (self.rank,destination,repr(data)))
24     return vector
```

Here is a revised version of the previous scalar product example using scatter:

Listing 8.7: in file: `psim_scalar2.py`

```
1  import random
2  from psim import PSim
3
4  def scalar_product_test2(n,p):
5      comm = PSim(p)
6      a = b = None
7      if comm.rank==0:
8          a = [random.random() for i in xrange(n)]
9          b = [random.random() for i in xrange(n)]
10     a = comm.one2all_scatter(0,a)
11     b = comm.one2all_scatter(0,b)
12
13     scalar = sum(a[i]*b[i] for i in xrange(len(a)))
14
15     scalar = comm.all2one_reduce(0,scalar)
16     if comm.rank == 0:
17         print scalar
18
19 scalar_product_test2(10,2)
```

8.3.3 Reduce

The all-to-one reduce pattern is very similar to the collect, except that the destination does not receive the entire list of values but some aggregated information about the values. The aggregation must be performed using a commutative binary function $f(x, y) = f(y, x)$. This guarantees that the reduction from the values go down in any order and thus are optimized for different network topologies.

The all-to-all reduce is similar to reduce, but every process will get the result of the reduction, not just one destination node. This may be achieved by an all-to-one reduce followed by a one-to-all broadcast:

Listing 8.8: in file: `psim.py`

```
def all2one_reduce(self,destination,value,op=lambda a,b:a+b):
    self.log("process %i: BEGIN all2one_reduce(%s)\n" % \
            (self.rank,repr(value)))
    self._send(destination,value)
    if self.rank==destination:
        result = reduce(op,[self._recv(i) for i in xrange(self.nprocs)])
    else:
        result = None
    self.log("process %i: END all2one_reduce(%s)\n" % \
            (self.rank,repr(value)))
    return result

def all2all_reduce(self,value,op=lambda a,b:a+b):
    self.log("process %i: BEGIN all2all_reduce(%s)\n" % \
            (self.rank,repr(value)))
    result=self.all2one_reduce(0,value,op)
    result=self.one2all_broadcast(0,result)
    self.log("process %i: END all2all_reduce(%s)\n" % \
            (self.rank,repr(value)))
    return result
```

And here are some examples of a reduce operation that can be passed to the `op` argument of the `all2one_reduce` and `all2all_reduce` methods:

Listing 8.9: in file: `psim.py`

```
@staticmethod
def sum(x,y): return x+y
@staticmethod
def mul(x,y): return x*y
@staticmethod
```

```
6      def max(x,y): return max(x,y)
7      @staticmethod
8      def min(x,y): return min(x,y)
```

Graph algorithms can also be parallelized, for example, the Prim algorithm. One way to do it is to represent the graph using an adjacency matrix where term i, j corresponds to the link between vertex i and vertex j. The term can be None if the link does not exist. Any graph algorithm, in some order, loops over the vertices and over the neighbors. This step can be parallelized by assigning different columns of the adjacency matrix to different computing processes. Each process only loops over some of the neighbors of the vertex being processed. Here is an example of the Prim algorithm:

Listing 8.10: in file: `psim_prim.py`

```
1   from psim import PSim
2   import random
3
4   def random_adjacency_matrix(n):
5       A = []
6       for r in range(n):
7           A.append([0]*n)
8       for r in range(n):
9           for c in range(0,r):
10              A[r][c] = A[c][r] = random.randint(1,100)
11      return A
12
13  class Vertex(object):
14      def __init__(self,path=[0,1,2]):
15          self.path = path
16
17  def weight(path=[0,1,2], adjacency=None):
18      return sum(adjacency[path[i-1]][path[i]] for i in range(1,len(path)))
19
20  def bb(adjacency,p=1):
21      n = len(adjacency)
22      comm = PSim(p)
23      Q = []
24      path = [0]
25      Q.append(Vertex(path))
26      bound = float('inf')
27      optimal = None
28      local_vertices = comm.one2all_scatter(0,range(n))
29      while True:
30          if comm.rank==0:
```

```
31        vertex = Q.pop() if Q else None
32    else:
33        vertex = None
34    vertex = comm.one2all_broadcast(0,vertex)
35    if vertex is None:
36        break
37    P = []
38    for k in local_vertices:
39        if not k in vertex.path:
40            new_path = vertex.path+[k]
41            new_path_length = weight(new_path,adjacency)
42            if new_path_length<bound:
43                if len(new_path)==n:
44                    new_path.append(new_path[0])
45                    new_path_length = weight(new_path,adjacency)
46                    if new_path_length<bound:
47                        bound = new_path_length # bcast
48                        optimal = new_path        # bcast
49                else:
50                    new_vertex = Vertex(new_path)
51                    P.append(new_vertex) # fix this
52                print new_path, new_path_length
53    x = (bound,optimal)
54    x = comm.all2all_reduce(x,lambda a,b: min(a,b))
55    (bound,optimal) = x
56
57    P = comm.all2one_collect(0,P)
58    if comm.rank==0:
59        for item in P:
60            Q+=item
61    return optimal, bound
62
63
64 m = random_adjacency_matrix(5)
65 print bb(m,p=2)
```

8.3.4 Barrier

Another global communication pattern is the barrier. It forces all processes when they reach the barrier to stop and wait until all the other processes have reached the barrier. Think of runners gathering at the starting line of a race; when all the runners are there, the race can start.

Here we implement it using a simple all-to-all broadcast:

Listing 8.11: in file: `psim.py`

```
 1    def barrier(self):
 2        self.log("process %i: BEGIN barrier()\n" % (self.rank))
 3        self.all2all_broadcast(0)
 4        self.log("process %i: END barrier()\n" % (self.rank))
 5        return
```

The use of `barrier` is usually a symptom of bad code because it forces parallel processes to wait for other processes without data actually being transferred.

8.3.5 Global running times

In the following table, we compute the order of growth or typical running times for the most common network topologies for typical communication algorithms:

Network	Send/Recv	One2All Bcast	Scatter
completely connected	1	1	1
switch	2	$\log p$	$2p$
1D mesh	$p - 1$	$p - 1$	p^2
2D mesh	$2(p^{\frac{1}{2}} - 1)$	$\sqrt{p}$	p^2
3D mesh	$3(p^{\frac{1}{3}} - 1)$	$p^{1/3}$	p^2
1D torus	$p/2$	$p/2$	p^2
2D torus	$2p^{\frac{1}{2}}$	$\sqrt{p}/2$	p^2
3D torus	$3/2p^{\frac{1}{3}}$	$p^{1/3}/2$	p^2
hypercube	$\log p$	$\log_2 p$	p
tree	$\log p$	$\log p$	p

It is obvious that the completely connected is the fastest network but also the most expensive to build. The tree is a cheap compromise. The switch tends to be faster for arbitrary point-to-point communication, but the switch comes to a premium. Multidimensional meshes and toruses become cost-effective when solving problems that are naturally defined on a grid because they only require next neighbor interaction.

8.4   mpi4py

The Psim emulator does not provide any actual speedup unless you have multiple cores or processors to execute the forked processes. A better approach would be to use mpi4py [66] because it allows running different processes on different machines on a network. mpi4py is a Python interface to the message passing interface (MPI). MPI is a standard protocol and API for interprocess communications. Its API are equivalent one by one to those of PSim, except that they have different names and different signatures.

Here is an example of using mpi4py:

```
1  from mpi4py import MPI
2
3  comm = MPI.COMM_WORLD
4  rank = comm.Get_rank()
5
6  if rank == 0:
7      message = "Hello World"
8      comm.send(message, dest=1, tag=11)
9  elif rank == 1:
10     message = comm.recv(source=0, tag=11)
11     print message
```

The comm object of class MPI.COMM_WORLD plays a similar role as the PSim object of the previous section. The MPI send and recv functions are very similar to the PSim equivalent ones, except that they require details about the type of the data being transferred and a communication tag. The tag allows node A to send multiple messages to B and allows B to receive them out of order. PSim does not allow tags.

8.5   Master-Worker and Map-Reduce

Map-Reduce [67] is a framework for processing highly distributable problems across huge data sets using a large number of computers (nodes). The group of computers is collectively referred to as a cluster (if all nodes use the same hardware) or a grid (if the nodes use different hardware). It comprises two steps:

"Map" (implemented here in a function mapfn): The master node takes the input data, partitions it into smaller subproblems, and distributes individual pieces of the data to worker nodes. A worker node may do this again in turn, leading to a multilevel tree structure. The worker node processes the smaller problem, computes a result, and passes that result back to its master node.

"Reduce" (implemented here in a function reducefn): The master node collects the partial results from all the subproblems and combines them in some way to compute the answer to the problem it needs.

Map-Reduce allows for distributed processing of the map and reduction operations. Provided each mapping operation is independent of the others, all maps can be performed in parallel—though in practice, it is limited by the number of independent data sources and/or the number of CPUs near each source. Similarly, a set of "reducers" can perform the reduction phase, provided all outputs of the map operation that share the same key are presented to the same reducer at the same time.

While this process can often appear inefficient compared to algorithms that are more sequential, Map-Reduce can be applied to significantly larger data sets than "commodity" servers can handle—a large server farm can use Map-Reduce to sort a petabyte of data in only a few hours, which would require much longer in a monolithic or single process system.

Parallelism also offers some possibility of recovering from partial failure of servers or storage during the operation: if one mapper or reducer fails, the work can be rescheduled—assuming the input data are still available.

Map-Reduce comprises of two main functions: mapfn and reducefn. mapfn takes a (key,value) pair of data with a type in one data domain and returns a list of (key,value) pairs in a different domain:

$$\text{mapfn}(k1, v1) \rightarrow (k2, v2) \tag{8.14}$$

The mapfn function is applied in parallel to every item in the input data set. This produces a list of (k2,v2) pairs for each call. After that, the Map-Reduce framework collects all pairs with the same key from all lists and groups them together, thus creating one group for each one of the different generated keys. The reducefn function is then applied in parallel to each group, which in turn produces a collection of values in the same domain:

$$reducefn(k2, [\text{list of } v2]) \rightarrow (k2, v3) \qquad (8.15)$$

The values returned by reducefn are then collected into a single list. Each call to reducefn can produce one, none, or multiple partial results. Thus the Map-Reduce framework transforms a list of (key, value) pairs into a list of values. It is necessary but not sufficient to have implementations of the map and reduce abstractions to implement Map-Reduce. Distributed implementations of Map-Reduce require a means of connecting the processes performing the mapfn and reducefn phases.

Here is a nonparallel implementation that explains the data workflow better:

```
 1  def mapreduce(mapper,reducer,data):
 2      """
 3      >>> def mapfn(x): return x%2, 1
 4      >>> def reducefn(key,values): return len(values)
 5      >>> data = xrange(100)
 6      >>> print mapreduce(mapfn,reducefn,data)
 7      {0: 50, 1: 50}
 8      """
 9      partials = {}
10      results = {}
11      for item in data:
12          key,value = mapper(item)
13          if not key in partials:
14              partials[key]=[value]
15          else:
16              partials[key].append(value)
17      for key,values in partials.items():
18          results[key] = reducer(key,values)
19      return results
```

And here is an example we can use to find how many random DNA strings contain the subsequence "ACTA":

```
1  >>> from random import choice
2  >>> strings = [''.join(choice('ATGC') for i in xrange(10))
3  ...                 for j in xrange(100)]
4  >>> def mapfn(string): return ('ACCA' in string, 1)
5  >>> def reducefn(check, values): return len(values)
```

```
6 >>> print mapreduce(mapfn,reducefn,strings)
7 {False: ..., True: ...}
```

The important thing about the preceding code is that there are two loops in Map-Reduce. Each loop consists of executing tasks (map tasks and reduce tasks) which are independent from each other (all the maps are independent, all the reduce are independent, but the reduce depend on the maps). Because they are independent, they can be executed in parallel and by different processes.

A simple and small library that implements the map-reduce algorithm in Python is mincemeat [68]. The workers connect and authenticate to the server using a password and request tasks to executed. The server accepts connections and assigns the map and reduce tasks to the workers.

The communication is performed using asynchronous sockets, which means neither workers nor the master is ever in a wait state. The code is event based, and communication only happens when a socket connecting the master to a worker is ready for a write (task assignment) or a read (task completed).

The code is also failsafe because if a worker closes the connection prematurely, the task is reassigned to another worker.

Function mincemeat uses the python libraries asyncore and asynchat to implement the communication patterns, for which we refer to the Python documentation.

Here is an example of a mincemeat program:

```
1 import mincemeat
2 from random import import choice
3
4 strings = [''.join(choice('ATGC') for i in xrange(10)) for j in xrange(100)]
5 def mapfn(k1, string): yield ('ACCA' in string, 1)
6 def reducefn(k2, values): return len(values)
7
8 s = mincemeat.Server()
9 s.mapfn = mapfn
10 s.reducefn = reducefn
11 s.datasource = dict(enumerate(strings))
12 results = s.run_server(password='changeme')
13 print results
```

Notice that in `mincemeat`, the data source is a list of key value dictionaries where the values are the ones to be processed. The key is also passed to the `mapfn` function as first argument. Moreover, the `mapfn` function can return more than one value using `yield`. This syntactical notation makes `minemeat` more flexible.

Execute this script on the server:

```
> python mincemeat_example.py
```

Run mincemeat.py as a worker on a client:

```
> python mincemeat.py -p changeme [server address]
```

You can run more than one worker, although for this example the server will terminate almost immediately.

Function `mincemeat` works fine for many applications, but sometimes one wishes for a more powerful tool that provides faster communications, support for arbitrary languages, and better scalability tools and monitoring tools. An example in Python is `disco`. A standard tool, written in Java but supporting Python, is **Hadoop**.

8.6   pyOpenCL

Nvidia should be credited for bringing GPU computing to the mass market. They have developed the CUDA [69] framework for GPU programming. CUDA programs consist of two parts: a host and a kernel. The host deploys the kernel on the available GPU core, and multiple copies of the kernel run in parallel.

Nvidia, AMD, Intel, and ARM have created the Kronos Group, and together they have developed the Open Common Language framework (OpenCL [70]), which borrows many ideas from CUDA and promises more portability. OpenCL supports Intel/AMD CPUs, Nvidia/ATI GPU, ARM chips, and the LLVM virtual machine.

OpenCL is a C99 dialect. In OpenCL, like in CUDA, there is a host program and a kernel. Multiple copies of the kernel are queued and run in parallel on available devices. Kernels running on the same device have access to a shared memory area as well as local memory areas.

A typical OpenCL program has the following structure:

```
1  find available devices (GPUs, CPUs)
2  copy data from host to device
3  run N copies of this kernel code on the device
4  copy data from device to host
```

Usually the kernel is written in C99, while the host is written in C++. It is also possible to write the host code in other languages, including Python. Here we will use the pyOpenCL [4] module for programming the host using Python. This produces no significative loss of performance compared to C++ because the actual computation is performed by kernel, not by the host. It is also possible to write the kernels using Python. This can be done using a library called Clyther [71] or one called ocl [5]. Here we will use the latter; ocl performs a one-time conversion of Python code for the kernel to C99 code. This conversion is done line by line and therefore also introduces no performance loss compared to writing native OpenCL kernels. It also provides an additional abstraction layer on top of pyOpenCL, which will make our examples more compact.

8.6.1 A first example with PyOpenCL

pyOpenCL uses numpy multidimensional arrays to store data. For example, here is a numpy example that performs the scalar product between two vectors, u and v:

```
1  import numpy as npy
2
3  size = 10000
4  u = npy.random.rand(size).astype(npy.float32)
5  v = npy.random.rand(size).astype(npy.float32)
6  w = npy.zeros(n,dtype=numpy.float32)
7
8  for i in xrange(0, n):
9      w[i] = u[i] + v[i];
10
11 assert npy.linalg.norm(w - (u + v)) == 0
```

The program works as follows:

- It creates a two numpy arrays u and v of given size and filled with random numbers.

- It creates another numpy array w of the same size filled with zeros.

- It loops over all indices of w and adds, term by term, u and v storing the result into w.

- It checks the result using the numpy linalg submodule.

Our goal is to parallelize the part of the computation performed in the loop. Notice that our parallelization will not make the code faster because this is a linear algorithm, and algorithms linear in the input are never faster when parallelized because the communication has the same order of growth as the algorithm itself:

```
from ocl import Device
import numpy as npy

n = 100000
u = npy.random.rand(n).astype(npy.float32)
v = npy.random.rand(n).astype(npy.float32)

device = Device()
u_buffer = device.buffer(source=a)
v_buffer = device.buffer(source=b)
w_buffer = device.buffer(size=b.nbytes)

kernels = device.compile("""
    __kernel void sum(__global const float *u, /* u_buffer */
                      __global const float *v, /* v_buffer */
                      __global float *w) {     /* w_buffer */
        int i = get_global_id(0);              /* thread id */
        w[i] = u[i] + v[i];
    }
    """)

kernels.sum(device.queue,[n],None,u_buffer,v_buffer,w_buffer)
w = device.retrieve(w_buffer)

assert npy.linalg.norm(w - (u+v)) == 0
```

This program performs the following steps in addition to the original non-OpenCL code: it declares a device object; it declares a buffer for each of the vectors u, v, and w; it declares and compiles the kernel; it runs the kernel; it retrieves the result.

The `device` object encapsulate the kernel(s) and a queue for kernel submission.

The line:

```
kernels.sum(...,[n],...)
```

submits to the queue n instances of the `sum` kernel. Each kernel instance can retrieve its own ID using the function `get_global_id(0)`. Notice that a kernel must be declared with the `__kernel` prefix. Arguments that are to be shared by all kernels must be `__global`.

The `Device` class is defined in the "ocl.py" file in terms of `pyOpenCL` API:

```
import numpy
import pyopencl as pcl

class Device(object):
    flags = pcl.mem_flags
    def __init__(self):
        self.ctx = pcl.create_some_context()
        self.queue = pcl.CommandQueue(self.ctx)
    def buffer(self,source=None,size=0,mode=pcl.mem_flags.READ_WRITE):
        if source is not None: mode = mode|pcl.mem_flags.COPY_HOST_PTR
        buffer = pcl.Buffer(self.ctx,mode, size=size, hostbuf=source)
        return buffer
    def retrieve(self,buffer,shape=None,dtype=numpy.float32):
        output = numpy.zeros(shape or buffer.size/4,dtype=dtype)
        pcl.enqueue_copy(self.queue, output, buffer)
        return output
    def compile(self,kernel):
        return pcl.Program(self.ctx,kernel).build()
```

Here `self.ctx` is the device context, `self.queue` is the device queue. The functions `buffer`, `retrieve`, and `compile` map onto the corresponding pyOpenCL functions `Buffer`, `enqueue_copy`, and `Program` but use a simpler syntax. For more details, we refer to the official `pyOpenCL` documentation.

8.6.2 Laplace solver

In this section we implement a two-dimensional Laplace solver. A three-dimensional generalization is straightforward. In particular, we want to solve the following differential equation known as a Laplace equation:

$$(\partial_x^2 + \partial_y^2)u(x,y) = q(x,y) \qquad (8.16)$$

Here q is the input and u is the output.

This equation originates, for example, in electrodynamics. In this case, q is the distribution of electric charge in space and u is the electrostatic potential.

As we did in chapter 3, we proceed by discretizing the derivatives:

$$\partial_x^2 u(x,y) = (u(x-h,y) - 2u(x,y) + u(x+h,y))/h^2 \qquad (8.17)$$
$$\partial_y^2 u(x,y) = (u(x,y-h) - 2u(x,y) + u(x,y+h))/h^2 \qquad (8.18)$$

Substitute them into eq. 8.16 and solve the equation in $u(x,y)$. We obtain

$$u(x,y) = 1/4(u(x-h,y) + u(x+h,y) + u(x,y-h) + u(x,y+h) - h^2 q(x,y)) \qquad (8.19)$$

We can therefore solve eq. 8.16 by iterating eq. 8.19 until convergence. The initial value of u will not affect the solution, but the closer we can pick it to the actual solution, the faster the convergence.

The procedure we utilized here for transforming a differential equation into an iterative procedure is a general one and applies to other differential equations as well. The iteration proceeds very much as the fixed point solver also examined in chapter 3.

Here is an implementation using `ocl`:

```
1  from ocl import Device
2  from canvas import Canvas
3  from random import randint, choice
4  import numpy
5
6  n = 300
7  q = numpy.zeros((n,n), dtype=numpy.float32)
8  u = numpy.zeros((n,n), dtype=numpy.float32)
9  w = numpy.zeros((n,n), dtype=numpy.float32)
10
```

```
11  for k in xrange(n):
12      q[randint(1, n-1),randint(1, n-1)] = choice((-1,+1))
13
14  device = Device()
15  q_buffer = device.buffer(source=q, mode=device.flags.READ_ONLY)
16  u_buffer = device.buffer(source=u)
17  w_buffer = device.buffer(source=w)
18
19
20  kernels = device.compile("""
21      __kernel void solve(__global float *w,
22                          __global const float *u,
23                          __global const float *q) {
24          int x = get_global_id(0);
25          int y = get_global_id(1);
26          int xy = y*WIDTH + x, up, down, left, right;
27          if(y!=0 && y!=WIDTH-1 && x!=0 && x!=WIDTH-1) {
28              up=xy+WIDTH; down=xy-WIDTH; left=xy-1; right=xy+1;
29              w[xy] = 1.0/4.0*(u[up]+u[down]+u[left]+u[right] - q[xy]);
30          }
31      }
32      """.replace('WIDTH',str(n)))
33
34  for k in xrange(1000):
35      kernels.solve(device.queue, [n,n], None, w_buffer, u_buffer, q_buffer)
36      (u_buffer, w_buffer) = (w_buffer, u_buffer)
37
38  u = device.retrieve(u_buffer,shape=(n,n))
39
40  Canvas().imshow(u).save(filename='plot.png')
```

We can now use the Python to C99 converter of ocl to write the kernel using Python:

```
1  from ocl import Device
2  from canvas import Canvas
3  from random import randint, choice
4  import numpy
5
6  n = 300
7  q = numpy.zeros((n,n), dtype=numpy.float32)
8  u = numpy.zeros((n,n), dtype=numpy.float32)
9  w = numpy.zeros((n,n), dtype=numpy.float32)
10
11  for k in xrange(n):
12      q[randint(1, n-1),randint(1, n-1)] = choice((-1,+1))
13
14  device = Device()
15  q_buffer = device.buffer(source=q, mode=device.flags.READ_ONLY)
```

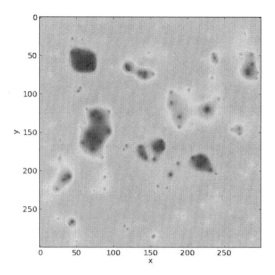

Figure 8.7: The image shows the output of the Laplace program and represents the two-dimensional electrostatic potential for a random charge distribution.

```
16  u_buffer = device.buffer(source=u)
17  w_buffer = device.buffer(source=w)
18
19  @device.compiler.define_kernel(
20      w='global:ptr_float',
21      u='global:const:ptr_float',
22      q='global:const:ptr_float')
23  def solve(w,u,q):
24      x = new_int(get_global_id(0))
25      y = new_int(get_global_id(1))
26      xy = new_int(x*n+y)
27      if y!=0 and y!=n-1 and x!=0 and x!=n-1:
28          up = new_int(xy-n)
29          down = new_int(xy+n)
30          left = new_int(xy-1)
31          right = new_int(xy+1)
32          w[xy] = 1.0/4*(u[up]+u[down]+u[left]+u[right] - q[xy])
33
34  kernels = device.compile(constants=dict(n=n))
35
36  for k in xrange(1000):
37      kernels.solve(device.queue, [n,n], None, w_buffer, u_buffer, q_buffer)
38      (u_buffer, w_buffer) = (w_buffer, u_buffer)
```

```
39
40 u = device.retrieve(u_buffer,shape=(n,n))
41
42 Canvas().imshow(u).save(filename='plot.png')
```

The output is shown in fig. 8.6.2.

One can pass constants to the kernel using

```
device.compile(..., constants = dict(n = n))
```

One can also pass include statements to the kernel:

```
device.compile(..., includes = ['#include <math.h>'])
```

where includes is a list of #include statements.

Notice how the kernel is line by line the same as the original C code. An important part of the new code is the define_kernel decorator. It tells ocl that the code must be translated to C99. It also declares the type of each argument, for example,

```
...define_kernel(... u='global:const:ptr_float' ...)
```

It means that:

```
global const float* u
```

Because in C, one must declare the type of each new variable, we must do the same in ocl. This is done using the pseudo-casting operators new_int, new_float, and so on. For example,

```
a = new_int(b+c)
```

is converted into

```
int a = b+c;
```

The converter also checks the types for consistency. The return type is determined automatically from the type of the object that is returned. Python objects that have no C99 equivalent like lists, tuples, dictionaries, and sets are not supported. Other types are converted based on the following table:

ocl	C99/OpenCL
a = new_*type*(...)	*type* a = ...;
a = new_prt_*type*(...)	*type* *a = ...;
a = new_prt_prt_*type*(...)	*type* **a = ...;
None	null
ADDR(x)	&x
REFD(x)	*x
CAST(prt_*type*,x)	(*type**)x

8.6.3   Portfolio optimization (in parallel)

In a previous chapter, we provided an algorithm for portfolio optimization. One critical step of that algorithm was the knowledge of all-to-all correlations among stocks. This step can efficiently be performed on a GPU.

In the following example, we solve the same problem again. For each time series k, we compute the arithmetic daily returns, r[k,t], and the average returns, mu[k]. We then compute the covariance matrix, cov[i,j], and the correlation matrix, cor[i,j]. We use different kernels for each part of the computation.

Finally, to make the application more practical, we use MPT [34] to compute a tangency portfolio that maximizes the Sharpe ratio under the assumption of Gaussian returns:

$$\max_x \frac{\mu^T x - r_{\text{free}}}{\sqrt{x^T \Sigma x}} \tag{8.20}$$

Here μ is the vector of average returns (mu), Σ is the covariance matrix (cov), and r_{free} is the input risk-free interest rate. The tangency portfolio is identified by the vector x (array x in the code) whose terms indicate the amount to be invested in each stock (must add up to $1). We perform this maximization on the CPU to demonstrate integration with the numpy linear algebra package.

We use the symbols i and j to identify the stock time series and the symbol t for time (for daily data t is a day); n is the number of stocks, and m is the number of trading days.

We use the `canvas` [11] library, based on the Python `matplotlib` library, to display one of the stock price series and the resulting correlation matrix. Following is the complete code. The output from the code can be seen in fig. 8.6.3.

```
from ocl import Device
from canvas import Canvas
import random
import numpy
from math import exp

n = 1000 # number of time series
m = 250  # number of trading days for time series
p = numpy.zeros((n,m), dtype=numpy.float32)
r = numpy.zeros((n,m), dtype=numpy.float32)
mu = numpy.zeros(n, dtype=numpy.float32)
cov = numpy.zeros((n,n), dtype=numpy.float32)
cor = numpy.zeros((n,n), dtype=numpy.float32)

for k in xrange(n):
    p[k,0] = 100.0
    for t in xrange(1,m):
        c = 1.0 if k==0 else (p[k-1,t]/p[k-1,t-1])
        p[k,t] = p[k,t-1]*exp(random.gauss(0.0001,0.10))*c

device = Device()
p_buffer = device.buffer(source=p, mode=device.flags.READ_ONLY)
r_buffer = device.buffer(source=r)
mu_buffer = device.buffer(source=mu)
cov_buffer = device.buffer(source=cov)
cor_buffer = device.buffer(source=cor)

@device.compiler.define_kernel(p='global:const:ptr_float',
                               r='global:ptr_float')
def compute_r(p, r):
    i = new_int(get_global_id(0))
    for t in xrange(0,m-1):
        r[i*m+t] = p[i*m+t+1]/p[i*m+t] - 1.0

@device.compiler.define_kernel(r='global:ptr_float',
                               mu='global:ptr_float')
def compute_mu(r, mu):
    i = new_int(get_global_id(0))
    sum = new_float(0.0)
```

```
40    for t in xrange(0,m-1):
41        sum = sum + r[i*m+t]
42    mu[i] = sum/(m-1)
43
44 @device.compiler.define_kernel(r='global:ptr_float',
45        mu='global:ptr_float', cov='global:ptr_float')
46 def compute_cov(r, mu, cov):
47    i = new_int(get_global_id(0))
48    j = new_int(get_global_id(1))
49    sum = new_float(0.0)
50    for t in xrange(0,m-1):
51        sum = sum + r[i*m+t]*r[j*m+t]
52    cov[i*n+j] = sum/(m-1)-mu[i]*mu[j]
53
54 @device.compiler.define_kernel(cov='global:ptr_float',
55                               cor='global:ptr_float')
56 def compute_cor(cov, cor):
57    i = new_int(get_global_id(0))
58    j = new_int(get_global_id(1))
59    cor[i*n+j] = cov[i*n+j] / sqrt(cov[i*n+i]*cov[j*n+j])
60
61 program = device.compile(constants=dict(n=n,m=m))
62
63 q = device.queue
64 program.compute_r(q, [n], None, p_buffer, r_buffer)
65 program.compute_mu(q, [n], None, r_buffer, mu_buffer)
66 program.compute_cov(q, [n,n], None, r_buffer, mu_buffer, cov_buffer)
67 program.compute_cor(q, [n,n], None, cov_buffer, cor_buffer)
68
69 r = device.retrieve(r_buffer,shape=(n,m))
70 mu = device.retrieve(mu_buffer,shape=(n,))
71 cov = device.retrieve(cov_buffer,shape=(n,n))
72 cor = device.retrieve(cor_buffer,shape=(n,n))
73
74 points = [(x,y) for (x,y) in enumerate(p[0])]
75 Canvas(title='Price').plot(points).save(filename='price.png')
76 Canvas(title='Correlations').imshow(cor).save(filename='cor.png')
77
78 rf = 0.05/m # input daily risk free interest rate
79 x = numpy.linalg.solve(cov,mu-rf) # cov*x = (mu-rf)
80 x *= 1.00/sum(x) # assumes 1.00 dollars in total investment
81 open('optimal_portfolio','w').write(repr(x))
```

Notice how the memory buffers are always one-dimensional, therefore the i,j indexes have to be mapped into a one-dimensional index i*n+j. Also notice that while kernels compute_r and compute_mu are called [n] times (once per stock k), kernels compute_cov and compute_cor are called [n,n] times, once per each couple of stocks `i,j`. The values of `i,j` are retrieved by `get_global_id(0)` and `(1)`, respectively.

In this program, we have defined multiple kernels and complied them at once. We call one kernel at the time to make sure that the call to the previous kernel is completed before running the next one.

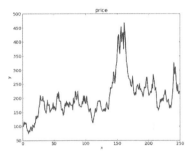 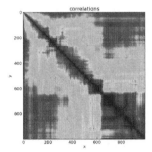

Figure 8.8: The image on the left shows one of the randomly generated stock price histories. The image on the right represents the computed correlation matrix. Rows and columns correspond to stock returns, and the color at the intersection is their correlation (red for high correlation and blue for no correlation). The resulting shape is an artifact of the algorithm used to generate random data.

9

Appendices

9.1 Appendix A: Math Review and Notation

9.1.1 Symbols

∞	infinity
∧	and
∨	or
∩	intersection
∪	union
∈	element or In
∀	for each
∃	exists
⇒	implies
:	such that
iff	if and only if

(9.1)

9.1.2 Set theory

Important sets

0	empty set	
$\mathbb{N}$	natural numbers $\{0,1,2,3,...\}$	
$\mathbb{N}^+$	positive natural numbers $\{1,2,3,...\}$	
$\mathbb{Z}$	all integers $\{...,-3,-2,-1,0,1,2,3,...\}$	(9.2)
$\mathbb{R}$	all real numbers	
$\mathbb{R}^+$	positive real numbers (not including 0)	
$\mathbb{R}^0$	positive numbers including 0	

Set operations

$\mathcal{A}$, $\mathcal{B}$ and $\mathcal{C}$ are some generic sets.

- **Intersection**

$$\mathcal{A} \cap \mathcal{B} \equiv \{x : x \in \mathcal{A} \text{ and } x \in \mathcal{B}\} \tag{9.3}$$

- **Union**

$$\mathcal{A} \cup \mathcal{B} \equiv \{x : x \in \mathcal{A} \text{ or } x \in \mathcal{B}\} \tag{9.4}$$

- **Difference**

$$\mathcal{A} - \mathcal{B} \equiv \{x : x \in \mathcal{A} \text{ and } x \notin \mathcal{B}\} \tag{9.5}$$

Set laws

- Empty set laws

$$\mathcal{A} \cup 0 = \mathcal{A} \tag{9.6}$$
$$\mathcal{A} \cap 0 = 0 \tag{9.7}$$

- Idempotency laws

$$\mathcal{A} \cup \mathcal{A} = \mathcal{A} \tag{9.8}$$
$$\mathcal{A} \cap \mathcal{A} = \mathcal{A} \tag{9.9}$$

- Commutative laws

$$A \cup B \ = \ B \cup A \tag{9.10}$$
$$A \cap B \ = \ B \cap A \tag{9.11}$$

- Associative laws

$$A \cup (B \cup C) \ = \ (A \cup B) \cup C \tag{9.12}$$
$$A \cap (B \cap C) \ = \ (A \cap B) \cap C \tag{9.13}$$

- Distributive laws

$$A \cap (B \cup C) \ = \ (A \cap B) \cup (A \cap C) \tag{9.14}$$
$$A \cup (B \cap C) \ = \ (A \cup B) \cap (A \cup C) \tag{9.15}$$

- Absorption laws

$$A \cap (A \cup B) \ = \ A \tag{9.16}$$
$$A \cup (A \cap B) \ = \ A \tag{9.17}$$

- DeMorgan laws

$$A - (B \cup C) \ = \ (A - B) \cap (A - C) \tag{9.18}$$
$$A - (B \cap C) \ = \ (A - B) \cup (A - C) \tag{9.19}$$

More set definitions

- A is a **subset** of B iff $\forall x \in A, x \in B$

- A is a **proper subset** of B iff $\forall x \in A, x \in B$ and $\exists x \in B, x \notin A$

- $P = \{S_i, i = 1, ..., N\}$ (a set of sets S_i) is a **partition** of A iff $S_1 \cup S_2 \cup ... \cup S_N = A$ and $\forall i, j, S_i \cap S_j = 0$

- The number of elements in a set A is called the **cardinality** of set A.

- cardinality($\mathbb{N}$)=countable infinite (∞)

- cardinality($\mathbb{R}$)=uncountable infinite (∞) !!!

Relations

- A **Cartesian Product** is defined as

$$\mathcal{A} \times \mathcal{B} = \{(a, b) : a \in \mathcal{A} \text{ and } b \in \mathcal{B}\} \qquad (9.20)$$

- A **binary relation** R between two sets $\mathcal{A}$ and $\mathcal{B}$ if a subset of their Cartesian product.

- A binary relation is **transitive** is aRb and bRc implies aRc

- A binary relation is **symmetric** if aRb implies bRa

- A binary relation is **reflexive** if aRa if always true for each a.

Examples:

- $a < b$ for $a \in \mathcal{A}$ and $b \in \mathcal{B}$ is a relation (transitive)

- $a > b$ for $a \in \mathcal{A}$ and $b \in \mathcal{B}$ is a relation (transitive)

- $a = b$ for $a \in \mathcal{A}$ and $b \in \mathcal{B}$ is a relation (transitive, symmetric and reflexive)

- $a \leq b$ for $a \in \mathcal{A}$ and $b \in \mathcal{B}$ is a relation (transitive, and reflexive)

- $a \geq b$ for $a \in \mathcal{A}$ and $b \in \mathcal{B}$ is a relation (transitive, and reflexive)

- A relation R that is transitive, symmetric and reflexive is called an **equivalence relation** and is often indicated with the notation $a \sim b$.

An equivalence relation is the same as a partition.

Functions

- A **function** between two sets $\mathcal{A}$ and $\mathcal{B}$ is a binary relation on $\mathcal{A} \times \mathcal{B}$ and is usually indicated with the notation $f : \mathcal{A} \longmapsto \mathcal{B}$

- The set $\mathcal{A}$ is called **domain** of the function.

- The set $\mathcal{B}$ is called **codomain** of the function.

- A function **maps** each element $x \in \mathcal{A}$ into an element $f(x) = y \in \mathcal{B}$

- The **image** of a function $f : \mathcal{A} \longmapsto \mathcal{B}$ is the set $\mathcal{B}' = \{y \in \mathcal{B} : \exists x \in \mathcal{A}, f(x) = y\} \subseteq \mathcal{B}$

- If $\mathcal{B}'$ is $\mathcal{B}$ then a function is said to be **surjective**.

- If for each x and x' in $\mathcal{A}$ where $x \neq x'$ implies that $f(x) \neq f(x')$ (e.g., if not two different elements of $\mathcal{A}$ are mapped into different element in $\mathcal{B}$) the function is said to be a **bijection**.

- A function $f : \mathcal{A} \longmapsto \mathcal{B}$ is **invertible** if it exists a function $g : \mathcal{B} \longmapsto \mathcal{A}$ such that for each $x \in \mathcal{A}, g(f(x)) = x$ and $y \in \mathcal{B}, f(g(y)) = y$. The function g is indicated with f^{-1}.

- A function $f : \mathcal{A} \longmapsto \mathcal{B}$ is a surjection and a bijection iff f is an invertible function.

Examples:

- $f(n) \equiv n \bmod 2$ with domain $\mathbb{N}$ and codomain $\mathbb{N}$ is not a surjection nor a bijection.

- $f(n) \equiv n \bmod 2$ with domain $\mathbb{N}$ and codomain $\{0, 1\}$ is a surjection but not a bijection

- $f(x) \equiv 2x$ with domain $\mathbb{N}$ and codomain $\mathbb{N}$ is not a surjection but is a bijection (in fact it is not invertible on odd numbers)

- $f(x) \equiv 2x$ with domain $\mathbb{R}$ and codomain $\mathbb{R}$ is not a surjection and is a bijection (in fact it is invertible)

-

9.1.3 Logarithms

If $x = a^y$ with $a > 0$, then $y = \log_a x$ with domain $x \in (0, \infty)$ and codomain $y = (-\infty, \infty)$. If the base a is not indicated, the natural log $a = e = 2.7183...$ is assumed.

Properties of logarithms:

$$\log_a x \;=\; \frac{\log x}{\log a} \tag{9.21}$$

$$\log xy \;=\; (\log x) + (\log y) \tag{9.22}$$

$$\log \frac{x}{y} \;=\; (\log x) - (\log y) \tag{9.23}$$

$$\log x^n \;=\; n \log x \tag{9.24}$$

9.1.4 Finite sums

Definition

$$\sum_{i=0}^{i<n} f(i) \equiv f(0) + f(1) + \ldots + f(n-1) \tag{9.25}$$

Properties

- **Linearity I**

$$\sum_{i=0}^{i\leq n} f(i) \;=\; \sum_{i=0}^{i<n} f(i) + f(n) \tag{9.26}$$

$$\sum_{i=a}^{i\leq b} f(i) \;=\; \sum_{i=0}^{i\leq b} f(i) - \sum_{i=0}^{i<a} f(i) \tag{9.27}$$

- **Linearity II**

$$\sum_{i=0}^{i<n} a f(i) + b g(i) = a \left(\sum_{i=0}^{i<n} f(i) \right) + b \left(\sum_{i=0}^{i<n} g(i) \right) \tag{9.28}$$

Proof:

$$
\begin{aligned}
\sum_{i=0}^{i<n} af(i) + bg(i) &= (af(0) + bg(0)) + \ldots + (af(n-1) + bg(n-1)) \\
&= af(0) + \ldots + af(n-1) + bg(0) + \ldots + bg(n-1) \\
&= a\left(f(0) + \ldots + f(n-1)\right) + b\left(g(0) + \ldots + g(n-1)\right) \\
&= a\left(\sum_{i=0}^{i<n} f(i)\right) + b\left(\sum_{i=0}^{i<n} g(i)\right) \quad (9.29)
\end{aligned}
$$

Examples:

$$
\sum_{i=0}^{i<n} c = cn \text{ for any constant } c \quad (9.30)
$$

$$
\sum_{i=0}^{i<n} i = \frac{1}{2}n(n-1) \quad (9.31)
$$

$$
\sum_{i=0}^{i<n} i^2 = \frac{1}{6}n(n-1)(2n-1) \quad (9.32)
$$

$$
\sum_{i=0}^{i<n} i^3 = \frac{1}{4}n^2(n-1)^2 \quad (9.33)
$$

$$
\sum_{i=0}^{i<n} x^i = \frac{x^n - 1}{x - 1} \text{ (geometric sum)} \quad (9.34)
$$

$$
\sum_{i=0}^{i<n} \frac{1}{i(i+1)} = 1 - \frac{1}{n} \text{ (telescopic sum)} \quad (9.35)
$$

9.1.5 Limits ($n \to \infty$)

In these section we will only deal with limits ($n \to \infty$) of positive functions.

$$
\lim_{n\to\infty} \frac{f(n)}{g(n)} = ? \quad (9.36)
$$

First compute limits of the numerator and denominator separately:

$$\lim_{n \to \infty} f(n) = a \qquad (9.37)$$

$$\lim_{n \to \infty} g(n) = b \qquad (9.38)$$

- If $a \in \mathbb{R}$ and $b \in \mathbb{R}^+$ then

$$\lim_{n \to \infty} \frac{f(n)}{g(n)} = \frac{a}{b} \qquad (9.39)$$

- If $a \in \mathbb{R}$ and $b = \infty$ then

$$\lim_{x \to \infty} \frac{f(x)}{g(x)} = 0 \qquad (9.40)$$

- If $(a \in \mathbb{R}^+$ and $b = 0)$ or $(a = \infty$ and $b \in \mathbb{R})$

$$\lim_{n \to \infty} \frac{f(n)}{g(n)} = \infty \qquad (9.41)$$

- If $(a = 0$ and $b = 0)$ or $(a = \infty$ and $b = \infty)$ use de l'Hopital rule

$$\lim_{n \to \infty} \frac{f(n)}{g(n)} = \lim_{n \to \infty} \frac{f'(n)}{g'(n)} \qquad (9.42)$$

and start again!

- Else ... the limit does not exist (typically oscillating functions or non-analytic functions).

For any $a \in \mathbb{R}$ or $a = \infty$

$$\lim_{n \to \infty} \frac{f(n)}{g(n)} = a \Rightarrow \lim_{n \to \infty} \frac{g(n)}{f(n)} = 1/a \qquad (9.43)$$

Table of derivatives

$f(x)$	$f'(x)$
c	0
ax^n	anx^{n-1}
$\log x$	$\frac{1}{x}$
e^x	e^x
a^x	$a^x \log a$
$x^n \log x, n > 0$	$x^{n-1}(n \log x + 1)$

(9.44)

Practical rules to compute derivatives

$$\frac{d}{dx}\left(f(x) + g(x)\right) = f'(x) + g'(x) \tag{9.45}$$

$$\frac{d}{dx}\left(f(x) - g(x)\right) = f'(x) - g'(x) \tag{9.46}$$

$$\frac{d}{dx}\left(f(x)g(x)\right) = f'(x)g(x) + f(x)g'(x) \tag{9.47}$$

$$\frac{d}{dx}\left(\frac{1}{f(x)}\right) = -\frac{f'(x)}{f(x)^2} \tag{9.48}$$

$$\frac{d}{dx}\left(\frac{f(x)}{g(x)}\right) = \frac{f'(x)}{g(x)} - \frac{f(x)g'(x)}{g(x)^2} \tag{9.49}$$

$$\frac{d}{dx}f(g(x)) = f'(g(x))g'(x) \tag{9.50}$$

Index

Bibliography

[1] http://www.python.org

[2] Travis Oliphant, "A Guide to NumPy". Vol.1. USA: Trelgol Publishing (2006)

[3] http://www.scipy.org/

[4] Andreas Klöckner *et al* "Pycuda and pyopencl: A scripting-based approach to gpu run-time code generation." Parallel Computing 38.3 (2012) pp 157-174

[5] https://github.com/mdipierro/ocl

[6] http://www.network-theory.co.uk/docs/pytut/

[7] http://oreilly.com/catalog/9780596158071

[8] http://www.python.org/doc/

[9] http://www.sqlite.org/

[10] http://matplotlib.sourceforge.net/

[11] https://github.com/mdipierro/canvas

[12] Stefan Behnel *et al.*, "Cython: The best of both worlds." Computing in Science & Engineering 13.2 (2011) pp 31-39

[13] Donald Knuth, "The Art of Computer Programming, Volume 3", Addison-Wesley, (1997). ISBN 0-201-89685-0

[14] Thomas H. Cormen, Charles E. Leiserson, Ronald L. Rivest, and

Clifford Stein, "Introduction to Algorithms", Second Edition. MIT Press and McGraw-Hill (2001). ISBN 0-262-03293-7

[15] J.W.J. Williams, J. W. J. "Algorithm 232 - Heapsort", Communications of the ACM 7 (6) (1964) pp 347–348

[16] E. F. Moore, "The shortest path through a maze", in Proceedings of the International Symposium on the Theory of Switching, Harvard University Press (1959) pp 285–292

[17] Charles Pierre Trémaux (1859–1882) École Polytechnique of Paris (1876). re-published in the Annals academic, March 2011 – ISSN: 0980-6032

[18] Joseph Kruskal, "On the Shortest Spanning Subtree of a Graph and the Traveling Salesman Problem", in Proceedings of the American Mathematical Society, Vol.7, N.1 (1956) pp 48–50

[19] R. C. Prim, "Shortest connection networks and some generalizations" in Bell System Technical Journal, 36 (1957) pp 1389–1401

[20] M. Farach-Colton et al., "Mathematical Support for Molecular Biology", DIMACS: Series in Discrete Mathematics and Theoretical Computer Science (1999) Volume 47. ISBN:0-8218-0826-5

[21] B. Korber et al., "Timing the Ancestor of the HIV-1 Pandemic Strains", Science (9 Jun 2000) Vol.288 no.5472.

[22] E. W. Dijkstra, "A note on two problems in connexion with graphs". Numerische Mathematik 1, 269–271 (1959). DOI:10.1007/BF01386390

[23] C. E. Shannon, "A Mathematical Theory of Communication". Bell System Technical Journal 27 (1948) pp 379–423

[24] R. M. Fano, "The transmission of information", Technical Report No. 65 MIT (1949)

[25] D. A. Huffman, "A Method for the Construction of Minimum-Redundancy Codes", Proceedings of the I.R.E., (1952) pp 1098–1102

[26] Bergroth and H. Hakonen and T. Raita, "A Survey of Longest Common Subsequence Algorithms". SPIRE (IEEE Computer Society) 39–48 (200). DOI:10.1109/SPIRE.2000.878178. ISBN:0-7695-0746-8

[27] Saul Needleman and Christian Wunsch, "A general method applicable to the search for similarities in the amino acid sequence of two proteins". Journal of Molecular Biology 48 (3) (1970) pp 443–53. DOI:10.1016/0022-2836(70)90057-4

[28] Tobias Dantzig, "Numbers: The Language of Science", 1930.

[29] V. Estivill-Castro, "Why so many clustering algorithms", ACM SIGKDD Explorations Newsletter 4 (2002) pp 65. DOI:10.1145/568574.568575

[30] J. J. Hopfield, "Neural networks and physical systems with emergent collective computational abilities", Proc. Natl. Acad. Sci. USA Vol. 79 (1982) pp 2554-2558

[31] Nils Aall Barricelli, Nils Aall, "Symbiogenetic evolution processes realized by artificial methods", Methodos (1957) pp 143–182

[32] Michael Garey and David Johnson, "Computers and Intractability: A Guide to the Theory of NP-Completeness", San Francisco: W. H. Freeman and Company (1979). ISBN:0-7167-1045-5

[33] Douglas R. Hofstadter, "Gödel, Escher, Bach: An Eternal Golden Braid", Basic Books 91979). ISBN:0-465-02656-7

[34] Harry Markowitz, "Foundations of portfolio theory", The Journal of Finance 46.2 (2012) pp 469-477

[35] P. E. Greenwood and M. S. Nikulin, "A guide to chi-squared testing". Wiley, New York (1996). ISBN:0-471-55779-X

[36] Andrew Lo and Jasmina Hasanhodzic, "The Evolution of Technical Analysis: Financial Prediction from Babylonian Tablets to Bloomberg Terminals", Bloomberg Press (2010). ISBN:1576603490

[37] Y. Saad and M.H. Schultz, "GMRES: A generalized minimal residual algorithm for solving nonsymmetric linear systems", SIAM J. Sci. Stat. Comput. 7 (1986). DOI:10.1137/0907058

[38] H. A. Van der Vorst, "Bi-CGSTAB: A Fast and Smoothly Converging Variant of Bi-CG for the Solution of Nonsymmetric Linear Systems". SIAM J. Sci. and Stat. Comput. 13 (2) (1992) pp 631–644. DOI:10.1137/0913035

[39] Richard Burden and Douglas Faires, "2.1 The Bisection Algorithm", Numerical Analysis (3rd ed.), PWS Publishers (1985). ISBN:0-87150-857-5

[40] Michiel Hazewinkel, "Newton method", Encyclopedia of Mathematics, Springer (2001). ISBN:978-1-55608-010-4

[41] Mordecai Avriel and Douglas Wilde, "Optimality proof for the symmetric Fibonacci search technique", Fibonacci Quarterly 4 (1966) pp 265–269 MR:0208812

[42] Loukas Grafakos, "Classical and Modern Fourier Analysis", Prentice-Hall (2004). ISBN:0-13-035399-X

[43] S.D. Poisson, "Probabilité des jugements en matière criminelle et en matière civile, précédées des règles générales du calcul des probabilitiés", Bachelier (1837)

[44] A. W. Van der Vaart, "Asymptotic statistics", Cambridge University Press (1998). ISBN:978-0-521-49603-2

[45] Jonah Lehrer, "How We Decide", Houghton Mifflin Harcourt (2009). ISBN:978-0-618-62011-1

[46] Edward N. Lorenz, "Deterministic non-periodic flow". Journal of the Atmospheric Sciences 20 (2) (1963) pp 130–141. DOI:10.1175/1520-0469

[47] Ian Hacking, "19th-century Cracks in the Concept of Determinism", Journal of the History of Ideas, 44 (3) (1983) pp 455-475 JSTOR:2709176

[48] F. Cannizzaro, G. Greco, S. Rizzo, E. Sinagra, "Results of the measurements carried out in order to verify the validity of the poisson-exponential distribution in radioactive decay events". The International Journal of Applied Radiation and Isotopes 29 (11) (1978) pp 649. DOI:10.1016/0020-708X(78)90101-1

[49] Yuval Perez, "Iterating Von Neumann's Procedure for Extracting Random Bits". The Annals of Statistics 20 (1) (1992) pp 590–597

[50] Martin Luescher, "A Portable High-Quality Random Number Generator for Lattice Field Theory Simulations", Comput. Phys. Commun. 79 (1994) pp 100-110

[51] http://demonstrations.wolfram.com/
PoorStatisticalQualitiesForTheRANDURandomNumberGenerator

[52] G. S. Fishman, "Grouping observations in digital simulation", Management Science 24 (1978) pp 510-521

[53] P. Good, "Introduction to Statistics Through Resampling Methods and R/S-PLUS", Wiley (2005). ISBN:0-471-71575-1

[54] J. Shao and D. Tu, "The Jackknife and Bootstrap", Springer-Verlag (1995)

[55] Paul Wilmott, "Paul Wilmott Introduces Quantitative Finance", Wiley 92005). ISBN:978-0-470-31958-1

[56] http://www.fas.org/sgp/othergov/doe/lanl/lib-www/la-pubs/00326407.pdf

[57] S. M. Ross, "Stochastic Processes", Wiley (1995). ISBN:978-0-471-12062-9

[58] Révész Pal, "Random walk in random and non random environments", World Scientific (1990)

[59] A.A. Markov. "Extension of the limit theorems of probability theory to a sum of variables connected in a chain". reprinted in Appendix B of R. Howard. "Dynamic Probabilistic Systems",

Vol.1, John Wiley and Sons (1971)

[60] W. Vervaat, "A relation between Brownian bridge and Brownian excursion". Ann. Prob. 7 (1) (1979) pp 143–149 JSTOR:2242845

[61] Kiyoshi Ito, "On stochastic differential equations", Memoirs, American Mathematical Society 4, 1–51 (1951)

[62] Steven Shreve, "Stochastic Calculus for Finance II: Continuous Time Models", Springer (2008) pp 114. ISBN:978-0-387-40101-0

[63] Sorin Istrail and Fumei Lam, "Combinatorial Algorithms for Protein Folding in Lattice Models: A Survey of Mathematical Results" (2009)

[64] Michael Flynn, "Some Computer Organizations and Their Effectiveness". IEEE Trans. Comput. C–21 (9) (1972) pp 948–960. DOI:10.1109/TC.1972.5009071

[65] Gene Amdahl, "Validity of the Single Processor Approach to Achieving Large-Scale Computing Capabilities", AFIPS Conference Proceedings (30) (1967) pp 483–485

[66] http://mpi4py.scipy.org/

[67] Jeffrey Dean and Sanjay Ghemawat. 2008. MapReduce: simplified data processing on large clusters. Commun. ACM 51, 1 (January 2008), 107-113. DOI=10.1145/1327452.1327492 http://doi.acm.org/10.1145/1327452.1327492

[68] http://remembersaurus.com/mincemeatpy/

[69] Erik Lindholm et al., "NVIDIA Tesla: A unified graphics and computing architecture." Micro, IEEE 28.2 (2008) pp 39-55.

[70] Aaftab Munshi, "OpenCL: Parallel Computing on the GPU and CPU." SIGGRAPH, Tutorial (2008).

[71] http://srossross.github.com/Clyther/

Made in the USA
San Bernardino, CA
28 April 2016